Use R!

Series Editors:
Robert Gentleman   Kurt Hornik   Giovanni Parmigiani

T0100708

Use R!

Deepayan Sarkar

Lattice

Multivariate Data Visualization with R

 Springer

Deepayan Sarkar
Program in Computational Biology
Division of Public Health Sciences
Fred Hutchinson Cancer Research Center
1100 Fairview Ave. N, M2-B876
Seattle, WA 98109-1024
USA
dsarkar@fhcrc.org

Series Editors:

Robert Gentleman
Program in Computational Biology
Division of Public Health Sciences
Fred Hutchinson Cancer Research Center
1100 Fairview Ave. N, M2-B876
Seattle, Washington 98109-1024
USA

Kurt Hornik
Department für Statistik und Mathematik
Wirtschaftsuniversität Wien Augasse 2-6
A-1090 Wien
Austria

Giovanni Parmigiani
The Sidney Kimmel Comprehensive Cancer Center at Johns Hopkins University
550 North Broadway
Baltimore, MD 21205-2011
USA

ISBN 978-0-387-75968-5   ISBN 978-0-387-75969-2 (eBook)
DOI 10.1007/978-0-387-75969-2

Library of Congress Control Number: 2008920682

Printed on acid-free paper.

9 8 7 6 5 4 3 2 1

springer.com

To Pinaki Mitra

Preface

The lattice package is software that extends the R language and environment for statistical computing (R Development Core Team, 2007) by providing a coherent set of tools to produce statistical graphics with an emphasis on multivariate data. It is modeled on the Trellis suite in S and S-PLUS®. From the user's point of view, it is a self-contained system that is largely independent of other graphics facilities in R. This book is about lattice, and is primarily intended for (1) both long-time and new R users looking for a powerful system to produce conventional statistical graphics, (2) existing lattice users willing to learn a little bit of R programming to gain increased flexibility, and (3) developers who wish to implement new graphical displays building on the infrastructure already available in lattice.

Why lattice?

Graphics can effectively complement statistical data analysis in various ways. Successful graphics arise from a combination of good design and good implementation. In this day and age, implementation is almost exclusively driven by computers. There is no lack of software tools that allow their users to convert data into graphics; lattice is yet another candidate in this ever-widening pool.

What makes lattice stand out? A good general-purpose tool should not get in the way of the user, yet it should be flexible enough to enable most tasks (without undue difficulty), whether it be standard, slightly out of the ordinary, or entirely novel. lattice tries to meet this standard by being a high-level tool that produces structured graphics, while retaining flexibility by systematically decoupling the various elements of a display; the individual elements have well thought-out defaults, but these can be overridden for detailed control. The end-product is a system that allows the creation of common statistical graphics, often with fairly complex structure, with very simple code. At the same time, it allows various degrees of customization, without requiring undue effort.

What to expect from this book

It is easy to get started with lattice, but the transition from seemingly simple to more sophisticated use can be difficult without an appreciation of how the different components and their defaults interact with each other. This appreciation can only come from experience, but it is hoped that this book can ease the transition to some extent.

The book started out as a manual for lattice, and was not intended to offer qualitative guidelines about the effective design of statistical graphics. This plan was abandoned quite early on; a static book is not the ideal vehicle for documenting an evolving system, and it is hard to look at and change bits and pieces of a picture without discussing its merits and drawbacks. In the end, this book consists of some comments on graphical design, some interesting (one would hope) examples, and large doses of lattice code and wisdom. It is still a book that is primarily about software; the code in the book is at least as important as the pictures. No code is hidden in this book, and if there is one key message that the reader should expect to take away, it is that lattice allows the creation of complex displays using relatively little code. This economy may not be appealing to everyone, but it is what I liked most about the Trellis system, and what has driven much of the development of lattice beyond the original goal of compatibility with Trellis. The other key idea that is difficult to communicate in function documentation, and one that is addressed in this book, is that of interrelationships between the different components of lattice, and how they can be effectively exploited.

What not to expect from this book

This book is not an exhaustive manual for lattice. Most functions in lattice are described to some extent in this book, but it does not serve as the definitive reference. There are two reasons for this. First, there are many features in lattice that are obscure and of very limited use, and do not justify detailed discussion. Second, lattice is an evolving system, and any attempt to document it exhaustively is sure to get out of date quickly. All functions in lattice come with online documentation, which should be used as the definitive reference.

How to read this book

That depends to a large extent on the reader. Those new to lattice should start with Chapter 1 to get a feel for what lattice is all about. Chapter 2 gives a more thorough, and sometimes quite technical, overview of the lattice model. Intermediate to advanced readers should find this chapter instructive. Beginners are encouraged to go through it as well, but should be prepared to encounter parts they find difficult, and skip them without getting bogged down; things should become clearer after gaining some practical experience.

The rest of Part I describes the various high-level functions in lattice. These chapters can be read in any order. Not much is said about the design of these graphics as they are standard, and most of the focus is on the software implementation. The level is basic for the most part; however, a few examples do go into some detail for the sake of taking a discussion to its natural conclusion. Again, beginners should be prepared to skip these parts during a first reading. Part II is more of a reference, going into the nitty-gritty details of lattice. A basic understanding of all the chapters is important to get the most out of lattice, but is not essential for casual use. These chapters too can be read in any order, for the most part, and the reader should feel free to pick and choose. The final two chapters, in Part III, deal with extensions to lattice, and are primarily intended for future developers. Of course, they can still be useful to the casual reader for the examples they provide.

It is important to realize that lattice is a complicated piece of software, and it is unrealistic to expect full mastery of it after one reading. The key to "getting it" is practical experience, and the best way to gain that experience is to try out the code. All the code in this book, along with the figures they produce, is available from the supporting Web site

http://lmdvr.r-forge.r-project.org/

A critical aspect of graphics that is hard to communicate in a book is its iterative nature; graphics that are presented to an audience is rarely the result of a first attempt. This process is reflected in some of the examples in this book, but many others have silently omitted many intermediate steps. One can get a sense of these missing steps by asking: "What is the purpose of this particular argument?" In other words, trying out variations of the code should be an integral part of the learning process.

The final thing to remember is that all this is the means to an end, namely, producing effective visualizations of data. Software can help, but the ultimate decisions are still the responsibility of the user. For those looking for guidance on how to create effective graphs, the work of Edward R. Tufte, William S. Cleveland, and of course John W. Tukey, are invaluable resources.

Color

Color can be an important factor in the visual impact of a graphic. Most figures in this book are black and white, but a few color plates are also included. Of these, some have the corresponding black and white versions as well, and have been chosen to highlight the impact of color. Others are solely available in color, as their black and white versions are of little or no use. Color versions of all figures are available on the book's Web site.

Prerequisites

No prior experience with lattice is required to read this book, but basic familiarity with R, and in particular the ability to use its online help system, is assumed. The first chapter of Dalgaard (2002) should suffice for the most part. Relatively advanced concepts such as generic functions and method dispatch are relevant, but can be ignored for casual use (these concepts are briefly introduced where relevant, but not at any deep level). No familiarity with traditional R graphics is presumed. Knowledge of the grid package can be beneficial, but is not essential.

Several R packages are used in this book. lattice itself should come with all recent installations of R, and it should be sufficient to type

```
> library("lattice")
```

at the R prompt to start using it. Other packages used explicitly (not counting further dependencies) are grid, latticeExtra, copula, ellipse, gridBase, flowViz, flowCore, hexbin, locfit, logspline, mapproj, maps, MASS, MEMSS, mlmRev, and RColorBrewer. All of these may not be of interest (some are required just for one or two examples); type

```
> help("install.packages")
```

to learn how to install the packages you need from CRAN.[1] flowCore, flowViz, and hexbin are Bioconductor packages, and may be installed by typing

```
> source("http://bioconductor.org/biocLite.R")
> biocLite(c("flowCore", "flowViz", "hexbin"))
```

A bit of history

The design of S graphics has been heavily influenced by the principles of graph construction laid out in *The Elements of Graphing Data* (Cleveland, 1985). This influence carries over to Trellis graphics, which incorporates further ideas (notably multipanel conditioning and banking) presented in *Visualizing Data* (Cleveland, 1993). Trellis graphics was first implemented in the S system, and has been available in S-PLUS for several years.

The name Trellis refers both to the general ideas underlying the system, as well as the specific implementation in S. The lattice package is an independent implementation of Trellis graphics (in the first sense), with an API closely modeled on the one in S. Unlike the S version, which is implemented using traditional graphics, lattice uses Paul Murrell's grid package, which provides more flexible low-level tools.

Although modeled on it, the lattice API is not identical to that of the Trellis suite in S. Some of the differences are due to the choice of grid as the underlying engine, but many are intentional. Still, where possible, effort has been made to allow Trellis code written in S to run with minimal modification. Consequently, writings about the original Trellis suite mostly apply to lattice as well. This includes the wealth of resources at the Trellis Web site at Bell Labs:

---

[1] The Comprehensive R Archive Network, http://cran.r-project.org

http://netlib.bell-labs.com/cm/ms/departments/sia/project/trellis/

However, the converse is not true. lattice has been extended beyond the original API in various ways, and is now at a point where it is difficult to partition its feature set into S-compatible ones and additional enhancements. For this reason, this book makes no attempt to distinguish between these, and presents Trellis graphics solely as implemented in the lattice package.

Caveats and alternatives

No system is perfect for all uses, and lattice is no exception. Trellis is a "high-level" paradigm by design, and lattice imposes considerable structure on the displays it creates. lattice allows a lot of wiggle room within these constraints while retaining its stylistic consistency and simple user interface, but this is not always enough. Fortunately, R provides excellent facilities for creating new displays from scratch, especially using the grid package. lattice itself is implemented using grid, and can benefit from the use of low-level facilities provided by it. Even for high-level graphics, R provides various alternatives. The traditional graphics system includes many high-level tools, which although not as proficient in dealing with multivariate data, often provide a richer set of options. Murrell (2005) gives a comprehensive overview of both traditional R graphics and grid graphics (as well as a brief introduction to lattice). The vcd package, inspired by Friendly (2000), provides many useful tools for categorical data, often with Trellis-style conditioning. Another high-level alternative is Hadley Wickham's ggplot2 (formerly ggplot) package, modeled on the approach of Wilkinson (1999), which is philosophically rather different from the Trellis approach. Like lattice, vcd and ggplot are also implemented using grid.

One thing R currently has virtually no support for is interactive graphics. Fortunately, some R packages provide interfaces to external systems that are better, notably rgl (OpenGL) and rggobi (GGobi). The playwith package written by Felix Andrews provides a modicum of interactivity within the R graphics framework, and works well with displays produced by lattice.

Acknowledgements

Ross Ihaka and Paul Murrell are the primary architects of the graphics infrastructure in R that lattice builds upon; lattice would have been particularly difficult to write without Paul's grid package. The lattice API is modeled on the original implementation of Trellis graphics in S. Implementing software is nowhere near as difficult as designing it, and the success of lattice owes much to the insight of the original designers. R is the wonderful platform that it is due to the efforts of its developer team and its vibrant user community, whose feedback has also driven much of the development of lattice beyond its original goals. My own introduction to R was eased by encouragement and guidance from Doug Bates and Saikat DebRoy.

Thanks to John Kimmel and Doug Bates for convincing me to start writing this book, and to Robert Gentleman for providing a wonderful environment in which to finish it. Several friends and colleagues commented on drafts of the book and the supporting Web site. Feedback from Springer's anonymous reviewers, and in particular John Maindonald, has been extremely valuable. On a technical note, this book was written on a number of different computers running Debian® GNU/Linux®, using the combination of Sweave and LaTeX as a document preparation system. The use of Emacs+ESS as the editing environment has improved productivity considerably.

Seattle, WA

December 2007   *Deepayan Sarkar*

Contents

Part III Extending Trellis Displays

1

Introduction

The traditional graphics subsystem in R is very flexible when it comes to producing standard statistical graphics. It provides a collection of high-level plotting functions that produce entire coherent displays, several low-level routines to enhance such displays and provide finer control over the various elements that make them up, and a system of parameters that allows global control over defaults and other details. However, this system is not very proficient at combining multiple plots in a page. It is quite straightforward to produce such plots; however, doing so in an effective manner, with properly coordinated scales, aspect ratios, and labels, is a fairly complex task that is difficult even for the experienced R user. Trellis graphics, originally implemented in S, was designed to address this shortcoming. The lattice add-on package provides similar capabilities for R users.

The name "Trellis" comes from the trellislike rectangular array of panels of which such displays often consist. Although Trellis graphics is typically associated with multiple panels, it is also possible to create single-panel Trellis displays, which look very much like traditional high-level R plots. There are subtle differences, however, mostly stemming from an important design goal of Trellis graphics, namely, to make optimum use of the available display area. Even single-panel Trellis displays are usually as good, if not better, than their traditional counterparts in terms of default choices. Overall, Trellis graphics is intended to be a more mature substitute for traditional statistical graphics in R. As such, this book assumes no prior knowledge of traditional R graphics; in fact, too much familiarity with it can be a hindrance, as some basic assumptions that are part and parcel of traditional R graphics may have to be unlearned. However, there are many parallels between the two: both provide high-level functions to produce comprehensive statistical graphs, both provide fine control over annotation and tools to augment displays, and both employ a system of user-modifiable global parameters that control the details of the display. This chapter gives a preview of Trellis graphics using a few examples; details follow in later chapters.

1.1 Multipanel conditioning

For the examples in this chapter, we make use of data on the 1997 A-level chemistry examination in Britain. The data are available in the mlmRev package, and can be loaded into R using the data() function.

```
> data(Chem97, package = "mlmRev")
```

A quick summary of the A-level test scores is given by their frequency table[1]

```
> xtabs(~ score, data = Chem97)
score
   0    2    4    6    8   10
3688 3627 4619 5739 6668 6681
```

Along with the test scores of 31,022 students, the dataset records their gender, age, and average GCSE score, which can be viewed as a pre-test achievement score. It additionally provides school and area-level information, which we ignore. In this chapter, we restrict ourselves to visualizations of the distribution of one continuous univariate measure, namely, the average GCSE score. We are interested in understanding the extent to which this measure can be used to predict the A-level chemistry examination score (which is a discrete grade with possible values 0, 2, 4, 6, 8, and 10).

1.1.1 A histogram for every group

One way to learn whether the final A-level score (the variable score) depends on the average GCSE score (gcsescore) is to ask a slightly different question: is the distribution of gcsescore different for different values of score? A popular plot used to summarize univariate distributions is the histogram. Using the lattice package, which needs to be attached first using the library() function, we can produce a histogram of gcsescore for each score, placing them all together on a single page, with the call

```
> library("lattice")
> histogram(~ gcsescore | factor(score), data = Chem97)
```

which produces Figure 1.1. There are several important choices made in the resulting display that merit attention. Each histogram occupies a small rectangular area known as a *panel*. The six panels are laid out in an array, whose dimensions are determined automatically. All panels share the same scales, which make the distributions easy to compare. Axes are annotated with tick marks and labels only along the boundaries, saving space between panels. A *strip* at the top of each panel describes which value of score that panel represents. These features are available in all Trellis displays, and are collectively

---

[1] Throughout this book, we make casual use of many R functions, such as data() and xtabs() here, without going into much detail. We expect readers to make use of R's online help system to learn more about functions that are unfamiliar to them.

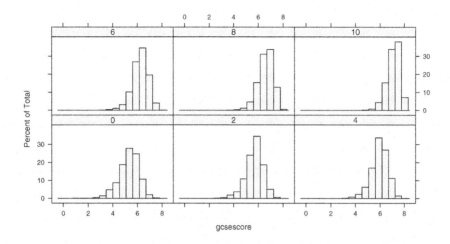

Figure 1.1. A conditional histogram using data on students attempting the A-level chemistry examination in Britain in 1997. The x-axis represents the average GCSE score of the students, and can be viewed as a prior achievement variable. The different panels represent subsets of students according to their grade in the A-level examination, and may be viewed as the response. Strips above each panel indicate the value of the response.

known as *multipanel conditioning*. These choices are intended to make the default display as useful as possible, but can be easily changed. Ultimate control rests in the hands of the user.

1.1.2 The Trellis call

Let us take a closer look at the `histogram()` call. As the name suggests, the `histogram()` function is meant to create histograms. In the call above, it has two arguments. The first (unnamed) argument, `x`, is a *"formula"* object that specifies the variables involved in the plot. The second argument, `data`, is a data frame that contains the variables referenced in the formula `x`.

The interpretation of the formula is discussed in more generality later, but is important enough to warrant some explanation here. In the formula

$$\text{~ gcsescore | factor(score)}$$

`factor(score)` (the part after the vertical bar symbol) is the *conditioning variable*, indicating that the resulting plot should contain one panel for each of its unique values (levels). The inline conversion to a factor is related to how the value of the conditioning variable is displayed in the strips; the reader is encouraged to see what happens when it is omitted. There can be more than one conditioning variable, or none at all.

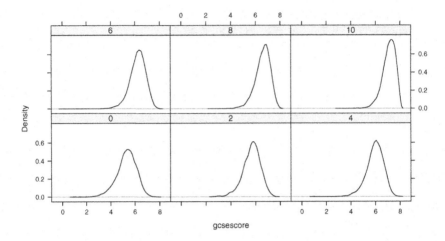

Figure 1.2. Conditional density plots. The data and design are the same as those in Figure 1.1, but histograms are replaced by kernel density estimates.

The part of the formula to the left of the conditioning symbol | specifies the *primary variable* that goes inside each panel; `gcsescore` in this case. What this part of the formula looks like depends on the function involved. All the examples we encounter in this chapter have the same form.

1.1.3 Kernel density plots

Histograms are crude examples of a more general class of univariate data summaries, namely, density estimates. `densityplot()`, another high-level function in the lattice package, can be used to graph kernel density estimates. A call that looks very much like the previous `histogram()` call produces Figure 1.2.

```
> densityplot(~ gcsescore | factor(score), data = Chem97,
              plot.points = FALSE, ref = TRUE)
```

There are two more arguments in this call: `ref`, which adds a reference line at 0, and `plot.points`, which controls whether in addition to the density, the original points will be plotted. Displaying the points can be informative for small datasets, but not here, with each panel having more than 3000 points. We show later that `ref` and `plot.points` are not really arguments of `densityplot()`, but rather of the default panel function, responsible for the actual plotting inside each panel.

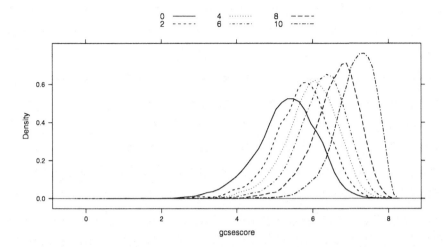

Figure 1.3. Grouped density plots. The density estimates seen in Figure 1.2 are now superposed within a single panel, forcing direct comparison. A legend on the top describes the association between levels of the grouping variable (**score** in this case) and the corresponding line parameters.

1.2 Superposition

Figures 1.1 and 1.2 both show that the distribution of **gcsescore** is generally higher for higher **score**. This pattern would be much easier to judge if the densities were superposed within the same panel. This is achieved by using **score** as a *grouping variable* instead of a conditioning variable[2] in the following call, producing Figure 1.3.

```
> densityplot(~ gcsescore, data = Chem97, groups = score,
              plot.points = FALSE, ref = TRUE,
              auto.key = list(columns = 3))
```

The **auto.key** argument automatically adds a suitable legend to the plot. Notice that it was not necessary to convert **score** into a factor beforehand; this conversion is done automatically. Another important point is that just as with variables in the formula, the expression specified as the **groups** argument was also evaluated in **Chem97** (the **data** argument). This is also true for another special argument, **subset**, which we learn about later.

An important theme in the examples we have seen thus far is the abstraction used in specifying the structure of a plot, which is essentially defined by the *type* of graphic (histogram, density plot) and the *role* of the variables involved (primary display, conditioning, superposition). This abstraction is fundamental in the lattice paradigm. Of course, calls as simple as these will not always suffice in real life, and lattice provides means to systematically

---

[2] This distinction between grouping and conditioning variables is specific to graphs.

control and customize the various elements that the graphic is comprised of, including axis annotation, labels, and graphical parameters such as color and line type. However, even when one ends up with a seemingly complex call, the basic abstraction will still be present; that final call will be typically arrived at by starting with a simple one and incrementally modifying it one piece at a time.

1.3 The *"trellis"* object

Most regular R functions do not produce any output themselves; instead, they return an object that can be assigned to a variable, used as arguments in other functions, and generally manipulated in various ways. Every such object has a class (sometimes implicit) that potentially determines the behavior of functions that act on them. A particularly important such function is the generic function `print()`, which displays any object in a suitable manner. The special property of `print()` is that it does not always have to be invoked explicitly; the result of an expression evaluated at the top level (i.e., not inside a function or loop), but not assigned to a variable, is printed automatically. Traditional graphics functions, however, are an exception to this paradigm. They typically do not return anything useful; they are invoked for the "side effect" of drawing on a suitable graphics device.

High-level functions in the lattice package differ in this respect from their traditional graphics analogues because they do not draw anything themselves; instead, they return an object, of class *"trellis"*. An actual graphic *is* created when such objects are "printed" by the `print()` method for objects of this class. The difference can be largely ignored, and lattice functions used just as their traditional counterparts (as we have been doing thus far), only because `print()` is usually invoked automatically. To appreciate this fact, consider the following sequence of commands.

```
> tp1 <- histogram(~ gcsescore | factor(score), data = Chem97)
> tp2 <-
      densityplot(~ gcsescore, data = Chem97, groups = score,
            plot.points = FALSE,
            auto.key = list(space = "right", title = "score"))
```

When these commands are executed, nothing gets plotted. In fact, `tp1` and `tp2` are now objects of class *"trellis"* that can, for instance, be summarized:

```
> class(tp2)
[1] "trellis"
> summary(tp1)
Call:
histogram(~gcsescore | factor(score), data = Chem97)

Number of observations:
factor(score)
```

```
   0    2    4    6    8   10
3688 3627 4619 5739 6668 6681
```

As noted above, the actual plots can be drawn by calling `print()`:

```
> print(tp1)
```

This may seem somewhat unintuitive, because `print()` normally produces text output in R, but it is necessary to take advantage of the automatic printing rule. The more natural

```
> plot(tp1)
```

has the same effect.

1.3.1 The missing Trellis display

Due to the automatic invocation of `print()`, lattice functions usually work as traditional graphics functions, where graphics output is generated when the user calls a function. Naturally, this similarity breaks down in contexts where automatic printing is suppressed. This happens, as we have seen, when the result of a lattice call is assigned to a variable. Unfortunately, it may also happen in other situations where the user may not be expecting it, for example, within `for()` or `while()` loops, or inside other functions. This includes the `source()` function, which is often used to execute an external R script, unless it is called with the `echo` argument set to TRUE. As with regular (non-graphics) R calls, the solution is to `print()` (or `plot()`) the result of the lattice call explicitly.

1.3.2 Arranging multiple Trellis plots

This object-based design has many useful implications, chief among them being the ability to arrange multiple lattice displays on a single page. Multipanel conditioning obviates the need for such usage to a large extent, but not entirely. For example, in Figure 1.4 we directly contrast the conditional histograms and the grouped density plots seen before. This is achieved by specifying the subregion to be occupied by a graphic on the fly when it is drawn, using optional arguments of the `plot()` method. Although this is one of the most common manipulations involving *"trellis"* objects explicitly, it is by no means the only one. A detailed discussion of *"trellis"* objects is given in Chapter 11.

1.4 Looking ahead

We have encountered two lattice functions in this chapter, `histogram()` and `densityplot()`. Each produces a particular type of statistical graphic, helpfully hinted at by its name. This sets the general trend: the lattice user interface principally consists of these and several other functions like these, each

```
> plot(tp1, split = c(1, 1, 1, 2))
> plot(tp2, split = c(1, 2, 1, 2), newpage = FALSE)
```

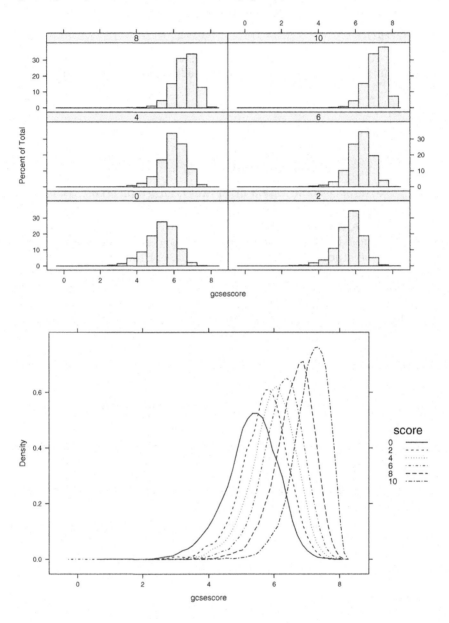

Figure 1.4. The conditional histogram and the grouped density plot of `gcsescore` by `score`, combined in a single figure. The comparison clearly illustrates the usefulness of superposition; the pattern of variance decreasing with mean that is obvious in the density plot is easy to miss in the histogram.

Function	Default Display
histogram()	Histogram
densityplot()	Kernel Density Plot
qqmath()	Theoretical Quantile Plot
qq()	Two-sample Quantile Plot
stripplot()	Stripchart (Comparative 1-D Scatter Plots)
bwplot()	Comparative Box-and-Whisker Plots
dotplot()	Cleveland Dot Plot
barchart()	Bar Plot
xyplot()	Scatter Plot
splom()	Scatter-Plot Matrix
contourplot()	Contour Plot of Surfaces
levelplot()	False Color Level Plot of Surfaces
wireframe()	Three-dimensional Perspective Plot of Surfaces
cloud()	Three-dimensional Scatter Plot
parallel()	Parallel Coordinates Plot

Table 1.1. High-level functions in the lattice package and their default displays.

intended to produce a particular type of graphic by default. The full list of high-level functions in lattice is given in Table 1.1. Chapters 3 through 6 focus on the capabilities of these high-level functions, describing each one in turn. The functions have much in common: they each have a formula interface that supports multipanel conditioning in a consistent manner, and respond to a number of common arguments. These common features, including the basics of multipanel conditioning, are described briefly in Chapter 2, and in further detail in Chapters 7 through 12. lattice is designed to be easily extensible using panel functions; some nontrivial examples are given in Chapter 13. Extensions can also be implemented as new high-level functions; Chapter 14 gives some examples and provides pointers for those who wish to create their own.

Part I

Basics

2

A Technical Overview of **lattice**

This chapter gives a broad overview of lattice, briefly describing the most important features shared by all high-level functions. Some of the topics covered are somewhat technical, but they are important motifs in the "big picture" view of lattice, and it would hinder rather than help to introduce them later at arbitrary points in the book. For readers that are new to lattice, it is recommended that they give this chapter a cursory overview and move on to the subsequent chapters. Each of the remaining chapters in Part I can be read, for the most part, directly after Chapter 1, although some advanced examples do require some groundwork laid out in this chapter. This nonlinear flow is inconvenient for those new to lattice, but it is somewhat inevitable; one should not expect to learn all the complexities of Trellis graphics in a first reading.

2.1 Basic usage

Strictly speaking, all high-level functions in lattice are generic functions, and suitable methods can be written for particular classes. In layman's terms, this means that the code that gets executed when a user calls such a function (e.g., `dotplot()`) will depend on the arguments supplied to the function. In practice, most such methods are simple wrappers to the *"formula"* method (i.e., the function that gets executed when the first argument is a *"formula"* object), because it allows for the most flexible specification of the structure of the display. Other methods can be valuable, as we see in later chapters. In this chapter, we restrict our attention to the *"formula"* methods.

2.1.1 The Trellis formula

The use of formulae is the standard when it comes to specifying statistical models in the S language, and they are the primary means of defining the structure of a lattice display as well. A typical Trellis formula looks like

$$y \sim x \mid a * b$$

The tilde (~) is what makes it a *"formula"* object, and is essential in any Trellis formula. Equally important is the vertical bar (|), which denotes conditioning. Variables (or more precisely, terms) to the right of the conditioning symbol are called *conditioning variables*, and those to the left are considered *primary variables*. A Trellis formula must contain at least one primary variable, but conditioning variables are optional. The conditioning symbol | must be omitted if there are no conditioning variables. There is no limit on the number of conditioning variables that can be specified, although the majority of actual use is covered by up to two. Conditioning variables may be separated by * or +; unlike many modeling functions, these are treated identically in lattice. The conditioning part of the formula has the same interpretation for all lattice functions, whereas that for the first part may vary by function. Thus,

$$\sim x$$

and

$$\log(z) \sim x * y \mid a + b + c$$

are both valid Trellis formulae (although not in all high-level functions). As the last example suggests, the formula can involve terms that are expressions involving one or more variables. After evaluation, all terms in the formula have to have the same length.

2.1.2 The data argument

Apart from specifying the structure of the display, use of a formula also allows one to separately specify, as the **data** argument, an object containing variables referenced in the formula. This is similar to other formula-based interfaces in R,[1] and reduces the temptation to use **attach()** (which is fraught with pitfalls) by obviating the need to repeatedly refer to the data source by name.[2] The **data** argument occupies the second position in the list of arguments in all high-level lattice functions, and is often not named in a call.

A less obvious implication of having a separate **data** argument is that methods can extend the types of objects that can act as a data source. The standard *"formula"* methods allow **data** to be data frames, lists, or environments (see **?eval**). In Chapter 14, we show how other types of objects may be used.

2.1.3 Conditioning

The case where the Trellis formula does not have any conditioning variables is fairly straightforward. To give analogies with base graphics functions, **histogram(~ x)** is similar to **hist(x)**, **xyplot(y ~ x)** is similar to **plot(x, y)**

---

[1] With the same caveats, briefly described in Section 10.1.

[2] An alternative is to use **with()**, which is sometimes more convenient.

or `plot(y ~ x)`, and so on. The rest of this chapter primarily deals with the situation where we do have one or more conditioning variables. In the first case, we can simply pretend to have one conditioning variable with a single level.

Conditioning variables are most often categorical variables, or *factors* in R parlance. They can also be *shingles*, which provide means to use continuous variables for conditioning.

Factors have a set of levels, representing its possible values. Each unique combination of the levels of the conditioning variables determines a *packet*, consisting of the subset of the primary variables that correspond to that combination.[3] It is possible for a packet to be empty if the corresponding combination of levels is not represented in the data. Each packet potentially provides the data for a single panel in the Trellis display, which consists of such panels laid out in an array of columns, rows, and pages. Choosing a proper layout is critical in obtaining an informative display; lattice tries to make the default choice as useful as possible, and provides ways to customize it.

Although packets are defined entirely by the formula, it is possible to omit or repeat certain levels of the conditioning variables when displaying a *"trellis"* object, in which case the corresponding packets may be omitted or repeated as well. Examples of these can be found in Figures 2.2 and 11.4.

2.1.4 Shingles

Multivariable relationships often involve many continuous variates, and the ability to condition on them is useful. Shingles afford a very general means to do so. The simplest possible approach to using a numeric variable for conditioning is to treat each of its unique values as a distinct level. This is, in fact, the default behavior in lattice. However, this is often unhelpful when the number of unique values is large. Another standard way to convert a continuous variate into an ordinal categorical variable is to discretize it, that is, to partition its range into two or more non-overlapping intervals, and replace each value by only an indicator of the interval to which it belonged. Such discretization can be performed by the R function `cut()`.

Shingles encompass both these ideas and extend them by allowing the intervals defining the discretization to overlap.[4] The intervals can be single points, or have no overlap, thus reducing to the two approaches described above. Each such interval is now considered a "level" of the shingle. Clearly, the level of a particular observation is no longer necessarily unique, as it can fall into more than one interval. This is not a hindrance to using the shingle

---

[3] Strictly speaking, a vector of subscripts indicating which rows in the original data contribute to the packet is also often part of the packet, although this is an irrelevant detail in most situations.

[4] Shingles are named after the overlapping pieces of wood or other building material often used to cover the roof or sides of a house.

as a conditioning variable; observations that belong to more than one level of the shingle are simply assigned to more than one packet.

This still leaves the issue of how best to choose the intervals that define a shingle, given a continuous variate. Cleveland (1993) suggests the "equal count" algorithm, which given a desired number of levels and amount of overlap, chooses the intervals so that each interval has roughly the same number of observations. This algorithm is used by the `equal.count()` function in the lattice package to create shingles from numeric variables. Shingles are discussed further in Chapter 10.

2.2 Dimension and physical layout

Multipanel conditioning can be viewed as an extended form of cross-tabulation, naturally conferring the concept of dimensions to *"trellis"* objects. Specifically, each conditioning variable defines a dimension, with extents given by the number of levels it has.

Consider the following graph, which uses data from a split-plot experiment (Yates, 1935) where yield of oats was measured for three varieties of oats and four nitrogen concentrations within each of six blocks.

```
> data(Oats, package = "MEMSS")
> tp1.oats <-
      xyplot(yield ~ nitro | Variety + Block, data = Oats, type = "o")
```

Although we do not formally encounter the `xyplot()` function until later, this is fairly typical usage, resulting in a scatter plot of yield against nitrogen concentration in each panel. The display produced by plotting `tp1.oats` is given in Figure 2.1. There are two conditioning variables (dimensions), with three and six levels. This is reflected in

```
> dim(tp1.oats)
[1] 3 6
> dimnames(tp1.oats)
$Variety
[1] "Golden Rain" "Marvellous"  "Victory"

$Block
[1] "I"   "II"  "III" "IV"  "V"   "VI"
```

These properties are shared by the cross-tabulation defining the conditioning.

```
> xtabs(~Variety + Block, data = Oats)
             Block
Variety       I II III IV V VI
  Golden Rain 4  4   4  4 4  4
  Marvellous  4  4   4  4 4  4
  Victory     4  4   4  4 4  4
```

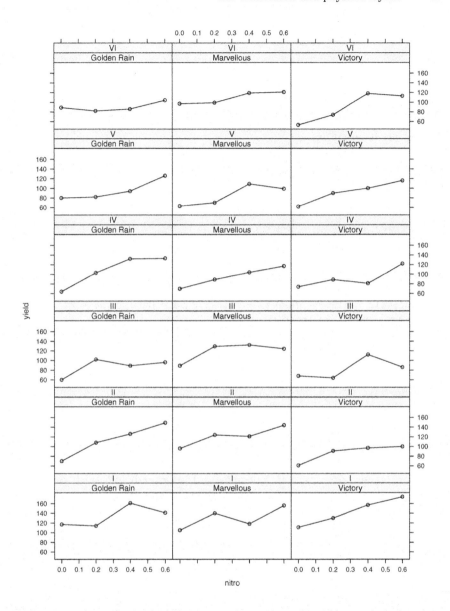

Figure 2.1. A Trellis display of the `Oats` data. The yield of oats is plotted against nitrogen concentration for three varieties of oats and six blocks. This is an example of a split-plot design. See `help(Oats, package = "MEMSS")` for more details about the experiment.

This cross-tabulation, with suitable modifications for shingles to account for the fact that packets may overlap, is in fact printed when a *"trellis"* object is summarized:

```
> summary(tp1.oats)
Call:
xyplot(yield ~ nitro | Variety + Block, data = Oats, type = "o")

Number of observations:
              Block
Variety       I II III IV V VI
  Golden Rain 4  4   4  4 4  4
  Marvellous  4  4   4  4 4  4
  Victory     4  4   4  4 4  4
```

It is possible to extract subsets of such objects in a natural way, treating them as arrays with the appropriate dimensions; for example,

```
> summary(tp1.oats[, 1])
Call:
xyplot(yield ~ nitro | Variety + Block, data = Oats, type = "o",
    index.cond = new.levs)

Number of observations:
              Block
Variety       I
  Golden Rain 4
  Marvellous  4
  Victory     4
```

The corresponding plot is shown in Figure 2.2. This view of a *"trellis"* object implies a linear ordering of the packets in it, similar to the ordering of elements in general arrays in R. Specifically, the order begins with the packet corresponding to the first index (level) of each dimension (conditioning variable) and proceeds by varying the index of the first dimension fastest, then the second, and so on. This order is referred to as the *packet order*.

Another array-like structure comes into play when a *"trellis"* object is actually displayed, namely, the physical layout of the panels. Whereas the number of dimensions of the abstract object is arbitrary, a display device is conventionally bound to two dimensions. Trellis displays, in particular, choose to divide the display area into a rectangular array of panels. An additional dimension is afforded by spreading out a display over multiple pages, which can be important in displays with a large number of combinations. All high-level lattice functions share a common paradigm that dictates how this layout is chosen, and provides common arguments to customize it to suit particular situations. Once the layout is determined, it defines the *panel order*, that is, the sequential order of panels in the three-way layout of columns, rows, and pages. The eventual display is created by matching packet order with panel

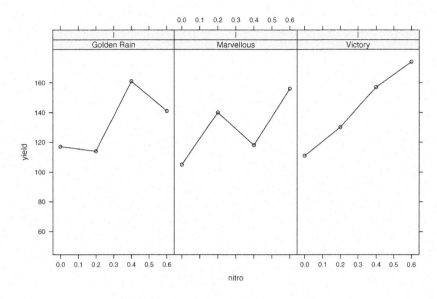

Figure 2.2. A Trellis display of a subset of a *"trellis"* object. The display represents the bottom row of Figure 2.1.

order.[5] The rest of this section discusses details of how the layout is controlled, and the choice of aspect ratio, which is closely related.

2.2.1 Aspect ratio

The aspect ratio of a panel is the ratio of its physical height and width. The choice of aspect ratio often plays a crucial role in determining the effectiveness of a display. There is no general prescription for choosing the aspect ratio, and one often needs to arrive at one by trial and error. In certain situations, a good aspect ratio can be automatically determined by the 45° banking rule, which is derived from the following idea. Consider a display, such as the `Oats` example above, where the changes in successive values (represented by line segments) contain information we wish to perceive. For a non-zero change, the corresponding line grows steeper as the aspect ratio increases, and shallower as it decreases. Cleveland et al. (1988) note that this information is best grasped when the orientation of such line segments is close to 45°, and recommend an algorithm that can be used to select an aspect ratio automatically based on this criterion. When the `aspect = "xy"` argument is specified in a high-level call, this 45° banking rule is used to compute the aspect ratio (see Chapter 8 for details). The `aspect` argument can also be an explicit numeric ratio, or the string `"iso"`, which indicates that the number of units per cm (i.e., the

---

[5] For the record, this can be changed; see `?packet.panel.default` for details.

relation between physical distance on the display and distance in the data scale) should be the same for both axes. This is appropriate in situations where the two scales have the same units, for example, in plots of spatial data, or plots of ROC curves where both axes represent probability.

2.2.2 Layout

A good choice of layout needs to take the aspect ratio into account. To make this point, let us look at Figure 2.3, which is produced by updating[6] Figure 2.1 to use an aspect ratio chosen by the 45° banking rule. As we can see, the default display does not make effective use of the available space. This is related to the rules that determine the default layout.

A Trellis display consists of several panels arranged in a rectangular array, possibly spanning multiple pages. The `layout` argument determines this arrangement. For an exact specification, `layout` should be a numeric vector giving the number of columns, rows, and pages in a multipanel display. Unless one wants to restrict the number of pages, the third element need not be specified; it is automatically chosen to accommodate all panels. The coordinate system used by default is like the Cartesian coordinate system: panels are drawn starting from the lower-left corner, proceeding first right and then up. This behavior can be changed by setting `as.table = TRUE` in a high-level lattice call,[7] in which case panels are drawn from the upper-left corner, going right and then down.

If there are two or more conditioning variables, `layout` defaults to the lengths of the first two dimensions, that is, the number of columns defaults to the number of levels of the first conditioning variable and the number of rows to the number of levels of the second conditioning variable (consequently, the number of pages is implicitly the product of the number of levels of the remaining conditioning variables, if any). This is clearly a sensible default, even though it is responsible for the somewhat awkward display in Figure 2.3.

The obvious way to "fix" Figure 2.3 is to switch the order of the conditioning variables. This can be done by regenerating the *"trellis"* object, or by simply transposing the existing one using

```
> t(tp1.oats)
```

However, we use another approach that makes use of a special form of the `layout` argument. The first element of `layout` can be 0, in which case its second element is interpreted as (a lower bound on) the total number of panels per page, leaving the software free to choose the exact layout. This is done by considering the aspect ratio and the device dimensions, and then choosing the layout so that the space occupied by each panel is maximized. The result of using this on our plot of the `Oats` data is given in Figure 2.4, where much better use is made of the available space.

---

[6] The `update()` function is formally discussed in Chapter 11.

[7] Chapter 7 describes how to change the default for `as.table` globally.

```
> update(tp1.oats,
          aspect="xy")
```

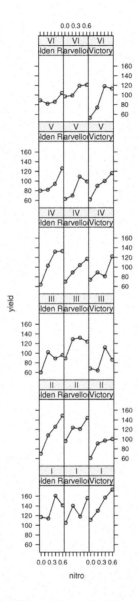

Figure 2.3. The display in Figure 2.1 updated to use the 45° banking rule to choose an aspect ratio. Although it is now easier to assess the changes in yield, the default layout results in considerable wastage of the available display area. This can be rectified using the **layout** argument, as we show in our next attempt.

```
> update(tp1.oats, aspect = "xy",
         layout = c(0, 18))
```

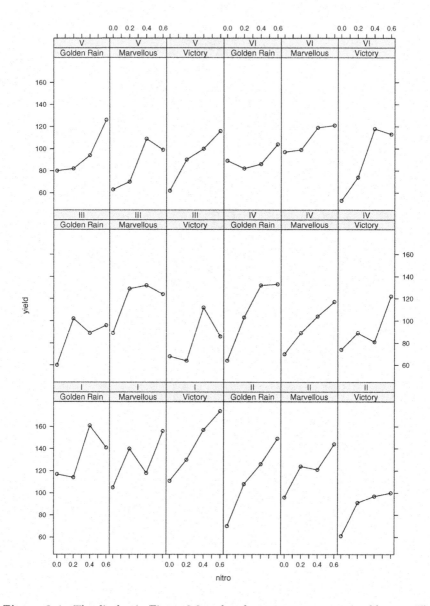

Figure 2.4. The display in Figure 2.3 updated to use an unconstrained layout. The aspect ratio calculated by the banking rule is taken into account when computing the layout, resulting in larger panels than before. However, there are now multiple blocks in each row of the layout, with no visual cue drawing attention to this fact.

```
> update(tp1.oats, aspect = "xy", layout = c(0, 18),
         between = list(x = c(0, 0, 0.5), y = 0.5))
```

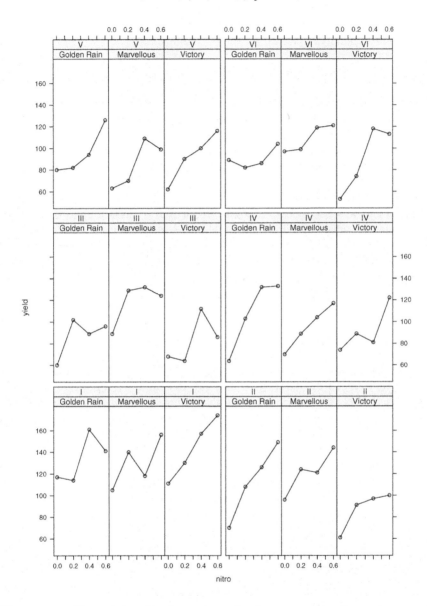

Figure 2.5. Figure 2.4 updated to put spacing between appropriate columns and rows, providing a visual cue separating panels into groups of blocks. This is possible because the layout happens to have exactly two blocks in every row; that is, none of the blocks spans multiple rows.

If there is only one conditioning variable with n levels, the default value of layout is c(0,n), thus taking advantage of this automatic layout computation. When aspect = "fill" (the default in most cases), this computation is carried out with an initial aspect ratio of 1, but in the eventual display the panels are expanded to fill up all the available space.

2.2.3 Fine-tuning the layout: between and skip

The between argument can be a list, with components x and y (both usually 0 by default) which are numeric vectors specifying the amount of blank space between the panels (in units of character heights). x and y are repeated to account for all panels in a page, and any extra components are ignored. This is often useful in providing a visual cue separating panels into blocks, as in Figure 2.5.

Another argument useful in fine-tuning the layout is skip, which is specified as a logical vector (default FALSE), replicated to be as long as the number of panels. For elements that are TRUE, the corresponding panel position is skipped; that is, nothing is plotted in that position. The panel that was supposed to be drawn there is now drawn in the next available panel position, and the positions of all the subsequent panels are bumped up accordingly. This is often useful for arranging plots in an informative manner.

2.3 Grouped displays

Trellis graphics is intended to foster easy and effective visualization of multivariate relationships in a dataset. As we saw in Chapter 1, a powerful construct that forces direct comparison is superposition, where data associated with different levels of a grouping variable are rendered together within a panel, but with different graphical characteristics. For example, different curves could be drawn in different color or line type, or points could be drawn with different symbols. Superposition is usually more effective than multipanel conditioning when the number of levels of the grouping variable is small. For many lattice functions, specifying a groups argument that refers to a categorical variable is enough to produce a "natural" grouped display.

We have seen grouped displays in Chapter 1. Perhaps the most well-known example in the context of Trellis graphics is Figure 1.1 from Cleveland (1993), which is recreated in Figure 2.6 using the following code.

```
> dotplot(variety ~ yield | site, barley,
          layout = c(1, 6), aspect = c(0.7),
          groups = year, auto.key = list(space = "right"))
```

The plot is a visualization of data from a barley experiment run in Minnesota in the 1930s (Fisher, 1971), and discussed extensively by Cleveland (1993). The plot effectively combines grouping and conditioning to highlight an anomaly in the data not easily noticed otherwise.

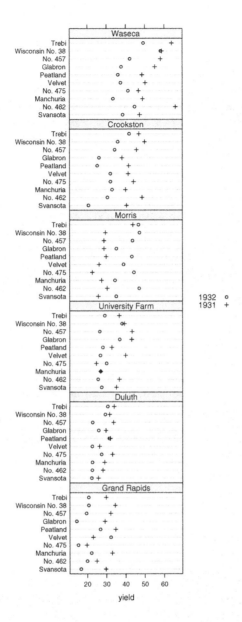

Figure 2.6. A multiway dot plot of data from a barley experiment run in Minnesota in the 1930s. Yield is plotted for several varieties of barley, conditioned on six sites. Different symbols are used to differentiate the year. The grouping and conditioning combine to highlight an anomaly in the data from Morris. Another subtle choice that enhances the effectiveness of the display is the ordering of the panels (sites) and the y variable (variety).

2.4 Annotation: Captions, labels, and legends

In Figure 2.6, as in Chapter 1, we have annotated the display by adding a legend, or key, that explains the correspondence of the different symbols to the respective levels of the grouping variable. Such legends are natural in grouped displays, but are not drawn by default. Usually, the simplest (although not the most general) way to add a suitable legend to a grouped display is to set draw.key = TRUE in the call. Often the key thus produced needs minor tinkering to get a more desirable result; this can be achieved by specifying auto.key as a list with suitable components. Generally speaking, legends can be placed in any of the four sides of a display, in which case enough space is automatically allocated for them. Alternatively, they can be placed anywhere inside the display, in which case no extra space is left, and the user has to make sure that they do not interfere with the actual display.

Other common means of annotating a display are to add meaningful captions and labels. Just as with traditional high-level graphics functions, most lattice functions allow the addition of four basic captions: a main title at the top (specified by the argument main), a subtitle at the bottom (sub), an x-axis label just below the x-axis (xlab), and a y-axis label to the left of the y-axis (ylab). xlab and ylab usually have some sensible defaults, whereas the other two are omitted. These labels are usually text strings, but can also be "*expression*" objects,[8] or for more generality, arbitrary grid objects (grobs). Another type of annotation directly supported by lattice functions is through the page argument. If specified, it has to be a function, and is called after each page is drawn. It can be used, for example, to mark the page numbers in a multipage display.

A full discussion of these annotation facilities is given in Chapter 9. Here, in Figure 2.7, we present one simple example with various labels and a legend. However, to fully appreciate even this simple example, we need to learn a little about how legends are specified.

2.4.1 More on legends

The construction of legends is a bit more involved than text labels, because they potentially have more structure. A template rich enough for most legends is one with (zero, one, or more) columns of text, points, lines, and rectangles, with suitably different symbols, colors, and so on. Such legends can be constructed using the draw.key() function, which can be indirectly used to add a legend to a plot simply by specifying a suitable list as the key argument in a high-level lattice function. To construct this list, we need to know what goes into the legend. The one in Figure 2.7 has a column of text with the levels of Variety, and a column of points with the corresponding symbols.

---

[8] Expressions, as typically produced by the expression() function, can be used to produce LATEX-like mathematical annotation, as described in the help page ?plotmath.

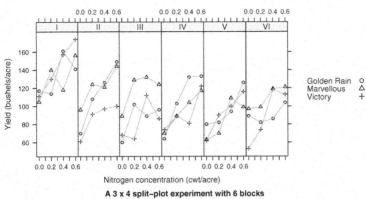

Figure 2.7. An alternative display of the `Oats` data. `Variety` is now used as a grouping variable, and a legend describes the association between its levels and the corresponding plotting characters. Various other labels are also included.

Here we run into a problem. The symbols and colors used by default in a lattice display are not determined until the plot is actually drawn, so that the current graphical settings can be taken into account (see Chapter 7 for details). For example, most plots on the pages of this book are black and white, but a reader trying to reproduce them will most likely do so interactively on a computer terminal, and will see them in color. In other words, when making the call to `xyplot()`, we do not know what the graphical parameters in the plot, and hence the legend, are going to be. A clumsy solution, used to produce Figure 2.7, is to bypass the problem by explicitly specifying the colors and symbols in the call itself.

```
> key.variety <-
      list(space = "right", text = list(levels(Oats$Variety)),
           points = list(pch = 1:3, col = "black"))
> xyplot(yield ~ nitro | Block, Oats, aspect = "xy", type = "o",
         groups = Variety, key = key.variety, lty = 1, pch = 1:3,
         col.line = "darkgrey", col.symbol = "black",
         xlab = "Nitrogen concentration (cwt/acre)",
         ylab = "Yield (bushels/acre)",
         main = "Yield of three varieties of oats",
         sub = "A 3 x 4 split-plot experiment with 6 blocks")
```

In most cases, a better solution is to use the `auto.key` argument, which we have already encountered on a couple of occasions. Chapter 9 examines this problem in more detail and explains the precise role of `auto.key`.

2.5 Graphing the data

At the end of the day, the usefulness of a statistical graphic is determined by how it renders the information it is supposed to convey. Multipanel conditioning, if used, imposes some preliminary structure on a Trellis display by systematically dividing up the data in a meaningful way. After determining these data subsets (packet) and their layout, they next need to be graphed. This involves a graphical encoding of the data, typically with a rendering of the relevant axes (tick marks and labels) to provide a frame of reference. For multipanel displays, an additional element describing each panel, specifically the associated levels of the conditioning variables, is necessary. This is done using strips, which can be customized or completely omitted by specifying a suitable `strip` (and in some cases `strip.left`) argument to any high-level lattice function (see Section 10.7 for details).

A fundamental assumption made in the Trellis design is that the nature of the graphical encoding will be repetitive; that is, the same *procedure* will be used to visualize each packet. This permits a decoupling of the procedures that draw the data and the axes, which can then be controlled separately. Recall that each panel in the display has an associated packet, a subset of the entire data. The exact form of a packet will depend on the high-level function used. Given the prescription for the graphic, a packet determines the data rectangle, a two-dimensional region enclosing the graphic. For example, in a bivariate scatter plot this is usually a rectangle defined by the range of the data; for a histogram, the horizontal extent of the data rectangle is the minimal interval containing all the bins, and the vertical scale ranges from 0 at the bottom to the height of the highest bin (which would depend on the type of histogram drawn) at the top. Another possibly relevant piece of information determined by the packet is a suitable aspect ratio for this data rectangle. In all lattice displays, these pieces of information are computed by the so-called prepanel function, which is discussed in detail in Chapter 8. Note that this view is not entirely satisfactory, as for some displays (e.g., scatter-plot matrices using `splom()` and three-dimensional scatter plots using `cloud()`) the usual axes have no meaning and the data display procedure itself has to deal with scales.

2.5.1 Scales and axes

For a single-panel display, one can proceed to draw the axes and the graphic once the data rectangle and aspect ratio are determined. However, for multipanel displays, there needs to be an intermediate step of combining the information from different packets. A common aspect ratio is chosen by some form of averaging if necessary. There are three alternative rules available to determine the scales. The default choice is to use the same data rectangle for each panel, namely, the smallest rectangle that encloses all individual data rectangles. This allows easy visual comparison between panels without constantly having to refer to the axes. This choice also allows the panels to share a

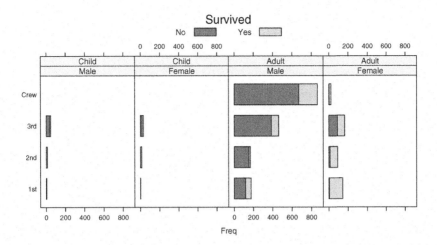

Figure 2.8. A bar chart summarizing the fate of passengers of the Titanic, classified by sex, age, and class. The plot is dominated by the third panel (adult males) as heights of the bars encode absolute counts, and all panels have the same limits.

common set of tick marks and axis labels along the boundary, saving valuable space. Sometimes this is not satisfactory because the ranges of the data in different packets are too different. If the data do not have a natural baseline and the relevant comparison is essentially done in terms of differences, it often suffices to have different scales as long as the number of units per cm is the same. The third choice, mainly useful for qualitative comparisons, is to allow completely independent scales, in which case the data rectangle for each panel is determined just by the corresponding packet. All these choices can be made selectively for either axis. The choice of which rule to use is controlled by the `scales` argument, which can also be used to control other aspects of axis annotation, such as the number of tick marks, position and labels of ticks, and so on. More directly, the arguments `xlim` and `ylim` allow explicit specification of the data rectangle, overriding the default calculations. This is an important and extensive topic, and is given due consideration in Chapter 8. We give one simple example here.

The `Titanic` dataset provides (as a four-dimensional array) a cross-tabulation of the fates of 2201 passengers of the famous ship, categorized by economic status (class), sex, and age. To use the data in a lattice plot, it is convenient to coerce it into a data frame. Our first attempt might look like the following, which produces Figure 2.8.

```
> barchart(Class ~ Freq | Sex + Age, data = as.data.frame(Titanic),
           groups = Survived, stack = TRUE, layout = c(4, 1),
           auto.key = list(title = "Survived", columns = 2))
```

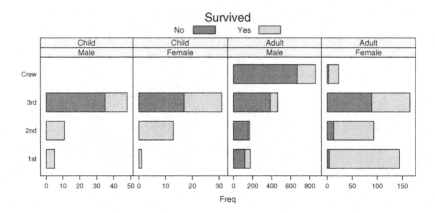

Figure 2.9. Survival among different subgroups of passengers on the Titanic, with a different horizontal scale in each panel. This emphasizes the proportion of survivors within each subgroup, rather than the absolute numbers. The proportion of survivors is smallest among third-class passengers, although the absolute number of survivors is not too low compared to the other classes.

All this plot really tells us is that there were many more males than females aboard (particularly among the crew, which is the largest group), and that there were even fewer children; which, although true, is unremarkable. The point we really want to make is that the "save women and children first" policy did not work as well for third-class passengers. This is more easily seen if we emphasize the proportions of survivors by allowing independent horizontal scales for different panels. Figure 2.9 is created using

```
> barchart(Class ~ Freq | Sex + Age, data = as.data.frame(Titanic),
           groups = Survived, stack = TRUE, layout = c(4, 1),
           auto.key = list(title = "Survived", columns = 2),
           scales = list(x = "free"))
```

2.5.2 The panel function

Once the rest of the structure (layout, data rectangles, annotation) is in place, packets are plotted in the appropriate panel. The actual plotting is done by a separate function, known as the *panel function* and specified as the `panel` argument, that is executed once for every panel with the associated data packet as its arguments. Each high-level lattice function has its own default panel function. By convention, the name of this function is given by "`panel.`" followed by the name of the high-level function. For example, the default panel function for `barchart()` is called `panel.barchart`, that for `histogram()` is `panel.histogram`, and so on. The remaining chapters in Part I describe the various high-level functions and their default panel functions in greater detail.

A lot can be achieved by the default panel functions, but one is not restricted to them by any means. In fact, it is the ability to define custom panel functions that allows the user to create novel Trellis displays easily, a process described in depth in Chapter 13. Even when predefined panel functions are adequate, an understanding of this process can greatly enhance the ability to use them effectively. For this reason, we spend some time here exploring this aspect. Readers new to R and lattice may want to skip the next part on first reading if they find it confusing.

2.5.3 The panel function demystified

Panel functions are, first and foremost, functions. This may sound obvious, but the concept of functions as arguments to other functions is often difficult to grasp for those not used to functional languages. To fix ideas, let us consider the call that produced Figure 2.9. As we plan to experiment just with the panel function, there is no point in repeating the full call every time. So, we save the object in a variable and use the update() method to manipulate it further.

```
> bc.titanic <-
      barchart(Class ~ Freq | Sex + Age, as.data.frame(Titanic),
               groups = Survived, stack = TRUE, layout = c(4, 1),
               auto.key = list(title = "Survived", columns = 2),
               scales = list(x = "free"))
```

Figure 2.9 can be reproduced by printing this object.

```
> bc.titanic
```

Because the default panel function for barchart() is panel.barchart(), this is equivalent to

```
> update(bc.titanic, panel = panel.barchart)
```

which has the same effect as specifying panel = panel.barchart in the original call. Note that the result of the call to update(), which is itself an object of class "trellis", has not been assigned to a variable and will thus be printed as usual. The variable bc.titanic remains unchanged. To make more explicit the notion that panel is a function, we can rewrite this as

```
> update(bc.titanic,
         panel = function(...) {
             panel.barchart(...)
         })
```

Although this does nothing new, it illustrates an important feature of the S language whose significance is easy for the beginner to miss; namely the ... argument. Complicated functions usually achieve their task by calling simpler functions. The ... argument in a function is a convenient way for it to capture arguments that are actually meant for another function called by it, without needing to know explicitly what those arguments might be. This trick

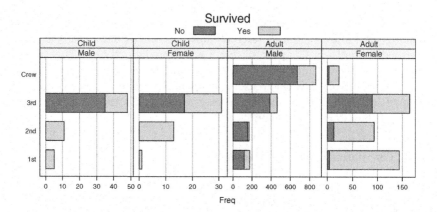

Figure 2.10. A modified version of Figure 2.9, with vertical reference lines added in the background. This is achieved using a custom panel function.

is very useful in lattice calls, because often one wants not to replace the panel function, but to add to it. A typical example is the addition of reference lines. The function **panel.grid()**, which is one of many utility functions in lattice, can be used to draw such reference lines as follows to produce Figure 2.10.

```
> update(bc.titanic,
         panel = function(...) {
             panel.grid(h = 0, v = -1)
             panel.barchart(...)
         })
```

Thanks to the ... argument, we used **panel.barchart()** without even knowing what arguments it accepts. It should also be noted that without the call to **panel.barchart()** in our custom panel function, only the reference lines would have been drawn.

Most default panel functions are designed to be quite flexible by themselves, and simple variations can frequently be achieved by changing one or more of their arguments. Suppose that we want to remove the black borders of the bars in Figure 2.10, which do not really serve any purpose as the bars are already shaded. Most panel functions have arguments to control the graphical parameters they use; in **panel.barchart()**, the border color is determined by the **border** argument (as described in the documentation). Thus, to make the borders transparent, we can use

```
> update(bc.titanic,
         panel = function(..., border) {
             panel.barchart(..., border = "transparent")
         })
```

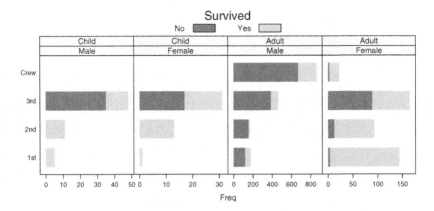

Figure 2.11. Another version of Figure 2.9 with the borders of the bars made transparent. This can be achieved using a custom panel function, but a simpler alternative is to specify a **border** argument to **barchart()** which is passed on to **panel.barchart()**.

which produces Figure 2.11. Once again, we make use of **panel.barchart()** without needing to know what its arguments are, except for the one we wanted to change.

This brings us to a simple, but extremely useful feature of high-level lattice functions. All of them have a ... argument, and will thus accept without complaint any number of extra named arguments. After processing the arguments it recognizes itself, a high-level function will collect all remaining arguments and pass them on to the panel function whenever it is called. The implication of this is that arguments that are intended for panel functions can be given directly to the high-level function. This panel function can of course be the default one, in which case the user does not even have to specify the panel function explicitly. Thus, an alternative way to produce Figure 2.11 is

```
> update(bc.titanic, border = "transparent")
```

We have already used this feature several times so far, and do so extensively in the next few chapters as well.

2.6 Return value

As briefly discussed in Chapter 1, high-level lattice functions do not draw anything themselves, instead returning an object of class *"trellis"*. In this chapter, we have made use of this fact several times without drawing attention to it, especially when calling the convenient **update()** method to make incremental changes to such objects. We learn more about *"trellis"* objects in Chapter 11.

3

Visualizing Univariate Distributions

Visualizing the distribution of a single continuous variable is a common graphical task for which several specialized methods have evolved. The distribution of a random variable X is defined by the corresponding cumulative distribution function (CDF) $F(x) = P(X \leq x)$. For continuous random variables, or more precisely, random variables with an absolutely continuous CDF, an equivalent representation is the density $f(x) = F'(x)$. One is often also interested in the inverse of F, the quantile function. R provides these functions for many standard distributions; for example, `pnorm()`, `dnorm()`, and `qnorm()` give the distribution, density, and quantile functions, respectively, for the normal distribution. Most of the visualization methods discussed in this chapter involve estimating these functions from data. In particular, density plots and histograms display estimates of the density f, and quantile plots and box-and-whisker plots are based on (partial) estimates of F or its inverse.

Although the mathematical relationships between the theoretical constructs are well-defined, there are no natural relationships between their standard estimates. Furthermore, the task of visualization comes with its own special rules; two plots with exactly the same information can put visual emphasis on entirely different aspects of that information. Thus, the appropriateness of a particular visualization depends to a large extent on the purpose of the analysis. We discuss the merits of different visualizations as we encounter them, but it is helpful to keep this background in mind when reading about them.

3.1 Density Plot

As we have already seen, using the `Chem97` dataset in Chapter 1, the `densityplot()` function produces kernel density plots. In that example, the densities estimated for the six `score` groups were all unimodal (i.e., they had one peak), and differed from each other essentially in their location, variability, and skewness (i.e., the first three moments). This is a common scenario in

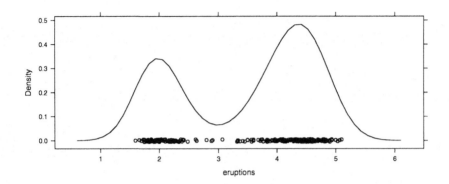

Figure 3.1. A kernel density plot of eruption times of the Old Faithful geyser. All optional arguments retain their defaults; in particular, the Gaussian kernel is used to compute the estimated density, and the raw data values are plotted with slight jittering.

many statistical analyses, but better graphical tools than density plots exist for it, as we show later in this chapter. Density plots are, however, particularly useful for detecting bimodality or multimodality.

Our first example uses the `faithful` dataset (Azzalini and Bowman, 1990; Härdle, 1990), a favorite in density estimation literature. The dataset records the duration of eruptions of the Old Faithful geyser in Yellowstone National Park, and the waiting time to the next eruption, over a period of a few days in 1985. We only look at the distribution of the durations. Figure 3.1 is produced by

```
> densityplot(~ eruptions, data = faithful)
```

By default, along with the estimated density, the points are plotted with some vertical jittering to address overplotting and ties. The `plot.points` argument can be used to change it to a "rug" as in the next example, or omit the points entirely.

There is a variety of approaches to density estimation, of which only the one implemented in the R function `density()` is available through the default panel function `panel.densityplot()`. It is fairly simple to implement other approaches, and we show an example in Figure 13.3. `density()` itself comes with several arguments to control the calculations, and these can be supplied directly to `densityplot()`. The two most important arguments are `kern`, which specifies the "kernel" used, and `bw`, which determines the bandwidth. The default kernel used in Figure 3.1 was the Gaussian. In Figure 3.2, we use the rectangular kernel instead, with a fixed bandwidth rather than a data-dependent one.

```
> densityplot(~ eruptions, data = faithful,
              kernel = "rect", bw = 0.2, plot.points = "rug", n = 200)
```

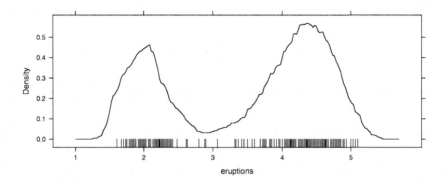

Figure 3.2. Another kernel density plot of the Old Faithful eruption times, this time using the rectangular kernel and a predetermined bandwidth. This is also known as an averaged shifted histogram (Scott, 1985), because it can be obtained as the average of all histograms with a fixed bin width.

Other kernel and bandwidth options are described in the help page for **density()**.

3.2 Large datasets

The datasets we have encountered so far are fairly small. Even the `Chem97` data, the largest we have seen, has only around 30,000 observations. Modern datasets, for example, those that arise from high-throughput biological assays, can easily exceed these sizes by many orders of magnitude. In this paradigm, careful thought is required about the storage of such data, as well as their analysis, including visualization. As a representative example, we use data from a flow cytometry (FCM) experiment. As we are primarily interested in issues related to visualization, we will use, for the most part, a small subset of the data that can be conveniently manipulated in the familiar data frame form. We briefly discuss the important practical issues of storage and efficiency in Chapter 14.

The full dataset (Rizzieri et al., 2007; Brinkman et al., 2007) originated from a collection of weekly peripheral blood samples obtained from several patients following allogeneic blood and marrow transplant. The goal of the study was to identify cellular markers that would predict the development of graft-versus-host disease (GvHD). Samples were taken at various time points before and after transplantation. Our "toy" example, available in the lattice-Extra package as the `gvhd10` dataset, represents samples obtained from one patient at seven time points. The blood samples were labeled with four different fluorescent probes to identify targeted biomarkers and a flow cytometer was used to determine fluorescent intensity for individual cells. The number

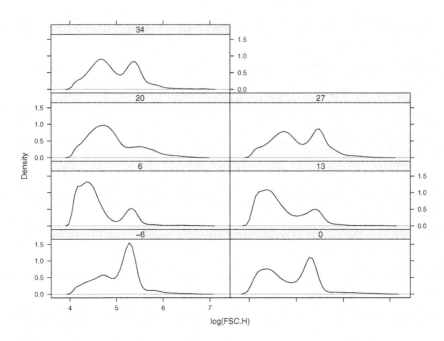

Figure 3.3. Kernel density plots of forward scatter (FSC) measurements of cells taken from a single transplant patient at different time points, on a logarithmic scale. The numbers in the strips represents days past transplant, with negative numbers representing days prior to it. FSC measurements are a surrogate for cell size.

of cells measured varied between approximately 4500 and 25,000 in the seven samples.

Flow cytometry data have their own special set of complexities that make analysis challenging. A full discussion of these complexities is beyond the scope of this book, but one important feature is that observations usually represent a mixture of multiple cell populations, not all of which are of interest. This is usually reflected in the marginal densities of individual marker intensities. In Figure 3.3, we look at densities of forward scatter, a measure of cell size, in blood samples taken at different time points.

```
> library("latticeExtra")
> data(gvhd10)
> densityplot(~log(FSC.H) | Days, data = gvhd10,
              plot.points = FALSE, ref = TRUE, layout = c(2, 4))
```

As the number of points is large, we do not plot them, instead adding a reference line. Other than this, the size of the dataset causes no problems in visualization, although the step of computing the density itself is more intensive.

3.3 Histograms

Histograms are also density estimates, somewhat cruder than kernel density estimates and possessing worse theoretical properties, but invaluable in the days before computers were ubiquitous. Histograms are created by dividing up the range of the data into non-overlapping bins, usually of the same length, and counting the number of observations that fall in them. Each bin is then represented by a rectangle with the bin as its base, where the height of the rectangle is computed to make its area equal the proportion of observations in that bin. This is formally known as the density histogram, because the result is a true probability density whose total area equals one. Other popular variants are the relative frequency histogram, where heights are relative frequencies, and the frequency histogram, where the heights are frequency counts within each bin. As long as all the bins have the same width, the heights in the three cases are multiples of each other (i.e., the corresponding histograms have the same shape but different y-scales). Unequal bin widths are rarely used outside introductory statistics textbooks. In Figure 3.4, we use the lattice function `histogram()` to present the same data as Figure 3.3. The `type` argument is used to compute the heights as density rather than the default of relative frequency. The number of bins (intervals) is increased to 50 because the number of observations is fairly large.

```
> histogram(~log2(FSC.H) | Days, gvhd10, xlab = "log Forward Scatter",
            type = "density", nint = 50, layout = c(2, 4))
```

This rendering emphasizes a feature not as obvious in the density plot, namely, that there is a fairly distinct lower bound for the observations, below which the density drops quite abruptly. This is an inherent limitation of kernel density estimates. Some alternative density estimation techniques can address this limitation if the bound is known in advance.

Despite this apparent advantage, it is rather difficult to justify the use of histograms in preference to density plots. For one thing, histograms are rather sensitive to the choice of bin locations; we would prefer estimates that depended more on the data and less on arbitrary parameter choices. Kernel density estimates can be viewed as a natural generalization of histograms that removes some of this arbitrariness, at least when the bins are of equal size. Specifically, consider a histogram with fixed bin width h. The histogram is entirely defined by the location of the left endpoint of any one bin, which is arbitrary. The *averaged shifted histogram* (ASH; Scott, 1985) removes this arbitrariness by defining the estimated density at a point x as the average value at x of all possible density histograms with bin width h. It can be easily shown that the estimate thus obtained is identical to the kernel density estimate computed using the rectangular kernel, as in Figure 3.2. Density plots are also preferable from the visualization perspective as they lend themselves more easily to superposition, as we have seen in Chapter 1. Histograms are nonetheless popular, not least because they are easier to explain to non-statisticians.

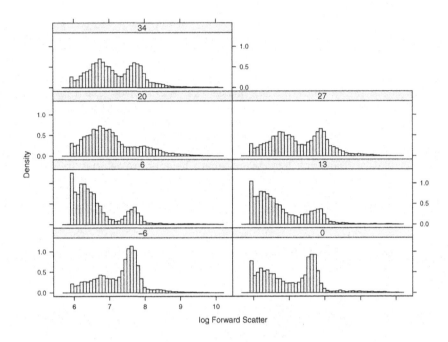

Figure 3.4. Histograms of log-transformed forward scatter measurements for different visits of a patient, with the same layout as Figure 3.3. A distinct lower bound for the measured values stands out much more clearly.

3.4 Normal Q–Q plots

A common task when analyzing continuous univariate data is to compare them to a theoretical distribution. Density estimates emphasize local features such as modes, but are not ideal for judging global features. The most commonly used tool for this job is the theoretical *quantile–quantile* (Q–Q) plot, which graphs quantiles of the observed data against similar quantiles of a probability distribution conjectured to be a reasonable match. For a good fit, a Q–Q plot is roughly linear, with systematic deviations suggesting a lack of fit. This is related to a well-known result from probability theory, that for a continuous random variable X with distribution function F, $F(X)$ has the uniform distribution $\mathcal{U}(0, 1)$, which in turn has a linear distribution function. Q–Q plots are particularly effective because the human eye finds it easier to perceive deviations from a straight line than from a curve.

We continue with the `Chem97` example from Chapter 1. The lattice function `qqmath()` can be used to create Q–Q plots comparing univariate data to a theoretical distribution. In principle, Q–Q plots can use any theoretical distribution. However, it is most common to use the normal distribution, which is the default choice in `qqmath()`. Figure 3.5 is produced by

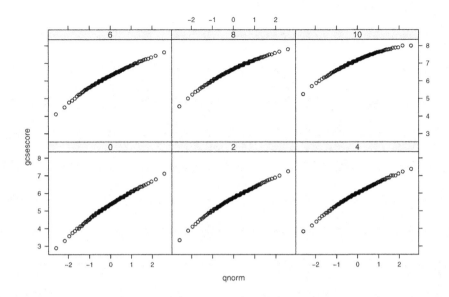

Figure 3.5. Normal Q–Q plots of average GCSE score for different final scores in the A-level chemistry exam. The systematic curvature in the Q–Q plot is indicative of a left-skewed distribution.

```
> qqmath(~ gcsescore | factor(score), data = Chem97,
        f.value = ppoints(100))
```

The formula and the **data** argument used should need no explanation. The other argument, **f.value**, tells qqmath() to use only 100 quantiles in each panel, instead of the default of as many quantiles as there are data points (which in this example would give more than 3000 points in each panel).

Figure 3.5 clearly shows systematic convexity, which is consistent with a left-skewed distribution. If we study the plot closely, we can confirm what we observed in Figure 1.3, that higher **score** is associated with higher **gcsescore**, and that the variance of **gcsescore** decreases with **score** (reflected in the decreasing slope of the Q–Q plots). This is clearer if we superpose the Q–Q plots in a single panel as we did with density plots in Chapter 1. Figure 3.6 is produced by

```
> qqmath(~ gcsescore | gender, Chem97, groups = score, aspect = "xy",
        f.value = ppoints(100), auto.key = list(space = "right"),
        xlab = "Standard Normal Quantiles",
        ylab = "Average GCSE Score")
```

We have also added **gender** as a conditioning variable and specified **aspect** = "xy", which chooses an aspect ratio using the 45° banking rule.

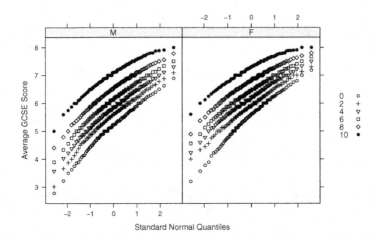

Figure 3.6. Normal Q–Q plots of average GCSE score by gender, grouped by final score. The aspect ratio has been chosen automatically using the 45° banking rule. The systematic curvature is still visible, and superposition now makes it easier to compare slopes, suggesting a systematic change in variance.

3.4.1 Normality and the Box–Cox transformation

The normal distribution plays an important role in many statistical analyses, and nice theoretical results follow if we can assume normality and equal variance, neither of which hold in our example. However, simple power transformations often improve the situation considerably. The Box–Cox transformation (Box and Cox, 1964) is a scale- and location-shifted version of the power transformation, given by

$$f_\lambda(x) = \frac{x^\lambda - 1}{\lambda}$$

for $\lambda \neq 0$, with $f_0(x) = \log x$. This formulation has the advantage of being continuous with respect to the "power" λ at $\lambda = 0$. The "optimal" Box–Cox transformation can be computed by the `boxcox()` function in the MASS package (Venables and Ripley, 2002). A plot of the profile log-likelihood function as a function of λ can be obtained using (result not shown)

```
> library("MASS")
> Chem97.pos <- subset(Chem97, gcsescore > 0)
> with(Chem97.pos,
       boxcox(gcsescore ~ score * gender, lambda = seq(0, 4, 1/10)))
```

One record with a `gcsescore` of 0 has to be omitted from the calculations. In this case, the optimal power is computed as $\lambda = 2.34$. We can visually confirm the success of this transformation using a Q–Q plot of the transformed values, shown in Figure 3.7.

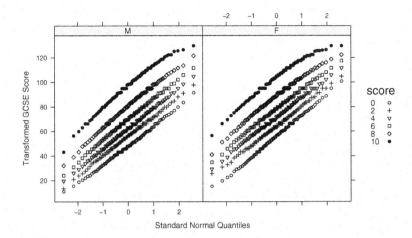

Figure 3.7. Normal Q–Q plots of transformed GCSE score. The transformation appears to have rectified most of the systematic departures from normality and homoscedasticity.

```
> Chem97.mod <- transform(Chem97, gcsescore.trans = gcsescore^2.34)
> qqmath(~ gcsescore.trans | gender, Chem97.mod, groups = score,
          f.value = ppoints(100), aspect = "xy",
          auto.key = list(space = "right", title = "score"),
          xlab = "Standard Normal Quantiles",
          ylab = "Transformed GCSE Score")
```

3.4.2 Other theoretical Q–Q plots

Although less common, distributions other than the normal can also be an appropriate choice as the source of the theoretical quantiles. For example, one standard choice is the uniform distribution, in which case the resulting Q–Q plot is related to the empirical distribution function of the data (see Figure 3.9). If the user is roughly familiar with the shape of common distribution functions, such plots can serve to suggest a good model for the data.

The primary use of quantile plots, however, is as a tool to compare two distributions, and its power comes from the fact that the human eye can better perceive deviations from a straight line than from a curve. To use this fact effectively, the two sets of quantiles compared using a Q–Q plot should arise from the same "expected" distribution. Viewed as a hypothesis test, this means that under the null hypothesis, the distributions compared are effectively the same (up to location and scale); a perceived departure from linearity in the Q–Q plot would lead to a rejection of the null hypothesis. Thus, if the data were expected to come from a certain distribution (not necessarily normal), it would be appropriate to compare against that distribution. One situation where this arises naturally is in simulation studies comparing the empirical and theoretical properties of the sampling distribution of some statistic. Such plots can also serve to demonstrate interesting properties of theoretical distributions; see Figure 10.5 for an example involving the exponential distribution.

3.5 The empirical CDF

A discussion of Q–Q plots would be incomplete without a mention of the empirical cumulative distribution function (ECDF). From a theoretical point of view, the ECDF is the non-parametric maximum likelihood estimate of the cumulative distribution function F. Trellis plots of the the ECDF can be produced by the ecdfplot() function in the latticeExtra package. Figure 3.8 is produced by

```
> library("latticeExtra")
> ecdfplot(~ gcsescore | factor(score), data = Chem97,
           groups = gender, auto.key = list(columns = 2),
           subset = gcsescore > 0, xlab = "Average GCSE Score")
```

The subset argument is used to remove a single outlier, shrinking the range of the data considerably.

Leaving aside certain technicalities that are largely irrelevant for visualization, the ECDF is closely related to a theoretical Q–Q plot with the uniform distribution as a reference, the difference being that the x- and y-axes are switched. An equivalent Q–Q plot, shown in Figure 3.9, is produced by

```
> qqmath(~ gcsescore | factor(score), data = Chem97, groups = gender,
         auto.key = list(points = FALSE, lines = TRUE, columns = 2),
         subset = gcsescore > 0, type = "l", distribution = qunif,
         prepanel = prepanel.qqmathline, aspect = "xy",
         xlab = "Standard Normal Quantiles",
         ylab = "Average GCSE Score")
```

It is easy to see that a normal Q–Q plot, or any other theoretical Q–Q plot for that matter, can be obtained by transforming the x-axis of a uniform Q–Q plot by a suitable theoretical quantile function. Similar transformations can be applied to an ECDF plot, but this is less common.

3.6 Two-sample Q–Q plots

Q–Q plots can also be used to directly compare two sets of observations. In theory, these are not much different from Q–Q plots against a theoretical distribution; quantiles from one sample are plotted not against corresponding quantiles from a theoretical distribution, but against those from the other sample. Two-sample Q–Q plots are are created by the qq() function. The formula defining such plots may seem somewhat unusual at first, but is natural when the data are stored in a single data frame, and extends naturally to the

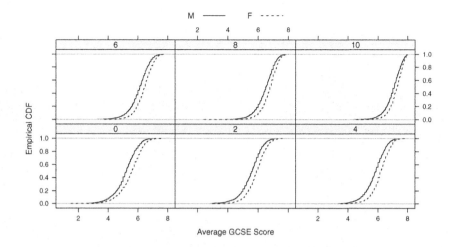

Figure 3.8. Empirical CDF plots of average GCSE scores by final score and gender. The empirical CDF is the non-parametric maximum likelihood estimate of the distribution function F.

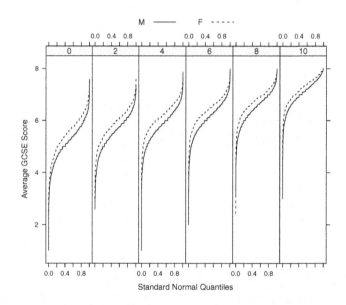

Figure 3.9. Uniform Q–Q plots of average GCSE scores. Modulo certain technicalities, these can be viewed as the inverse of ECDF plots, with the x- and y-axes switched. Neither are particularly useful as diagnostics for lack of fit, but can be used for comparing multiple distributions.

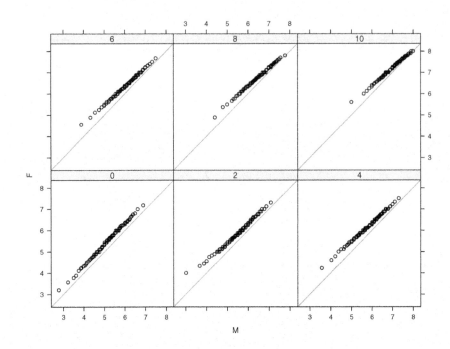

Figure 3.10. Two sample Q–Q plots comparing average GCSE score by gender (which conveniently has two levels), after conditioning on final score. The Q–Q plots are linear, but fall slightly above the diagonal and have slope less than 1 (except for the first panel), suggesting that the distributions of GCSE score are similar, with a higher mean and lower variance for females compared to males. An overall upward shift across panels is also apparent.

bwplot() function, which we encounter soon. Specifically, the formula has the form y ~ x, where x is a numeric vector that consists of both samples, and y is a factor of the same length as x with exactly two levels defining the two samples. Figure 3.10 shows a Q–Q plot comparing **gcsescore** for males and females after conditioning on **score**.

```
> qq(gender ~ gcsescore | factor(score), Chem97,
      f.value = ppoints(100), aspect = 1)
```

The two axes correspond to quantiles of the two samples. By default, both axes have the same limits, and a diagonal line is added for reference. In this case, the scatter in each panel is linear, but above the diagonal and not quite parallel to it. This suggests that the distributions are similar except for a scale and location change, with **gcsescore** values for females being slightly higher and less variable given a final **score**. A useful variant of this plot can be seen in Figure 11.5.

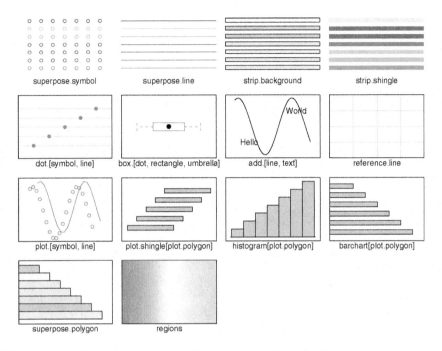

Figure 7.4. Color version. A visual summary of the default color parameter settings, produced by `show.settings()`.

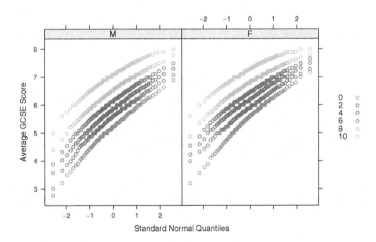

Figure 3.6. Color version. Grouping by the final A-level chemistry score is indicated by color, as opposed to plotting character in the black and white version. This makes the groups much easier to distinguish.

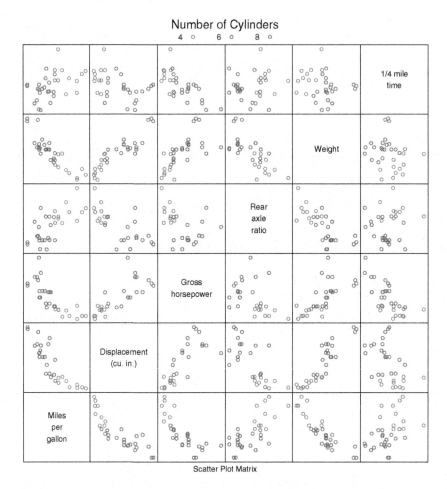

Figure 5.17. Color version. As with Figure 3.6, the use of color makes the grouping by number of cylinders visually much more prominent.

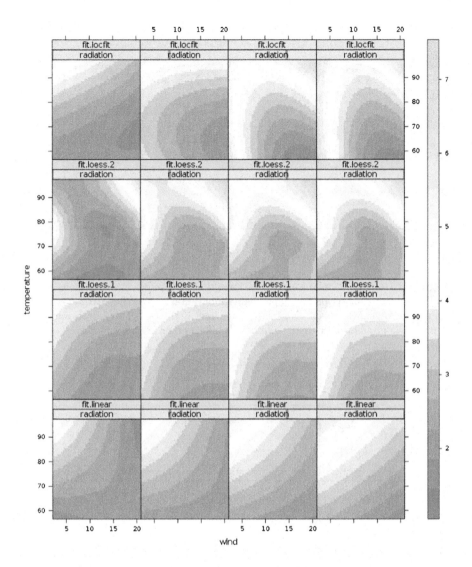

Figure 6.9. Color version. Black and white level plots can only show one gradient, whereas true color allows more choices. Here, the middle of the range is encoded by white (0 saturation), with gradually strengthening saturation towards the cyan and magenta hues away from the middle (Cleveland, 1993).

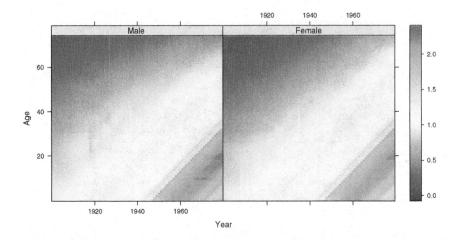

Figure 6.19. A level plot showing estimated U.S. population (in millions) by age and year, conditioning on sex. The "baby boom" starting in the late 1940s is quite prominent. A subtle effect seen in the left panel is the temporary drop among the young male population in 1918; whether the effect is noticeable depends strongly on the color scheme used.

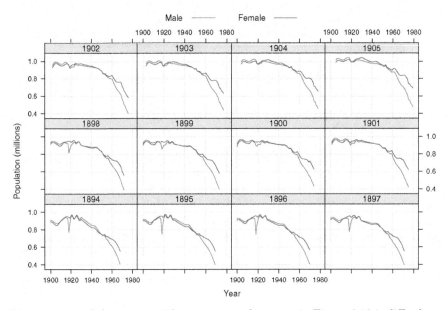

Figure 10.9. Color version. The temporary drop seen in Figure 6.19 is difficult to miss in this display. Encoding groups by color rather than line type, as in the black and white version, makes comparison easier and reduces the artificial prominence given to solid lines over broken line types.

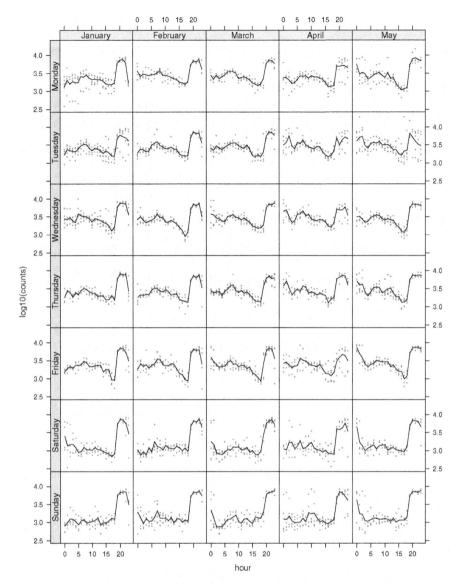

Figure 11.6. Color version. The use of color allows us to put different emphasis on the lines and the points.

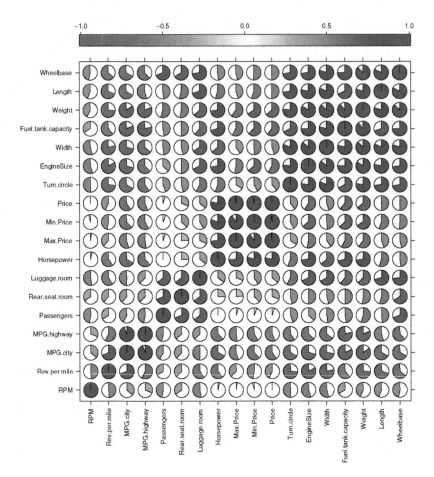

Figure 13.6. A corrgram as described by Friendly (2002), showing a correlation matrix derived from the `Cars93` data. In addition to color, correlations are encoded using the amount of fill-in in circular "Pac-man" symbols.

American
English
French (except Basque)
German
Irish
Italian
Norwegian
Other

Figure 13.8. Modal ancestry (ancestry with the highest frequency) in the U.S. 2000 census, by county. No projection scheme is used, resulting in a somewhat odd-looking display.

American   French (except Basque)   Irish   Norwegian
English   German   Italian   Other

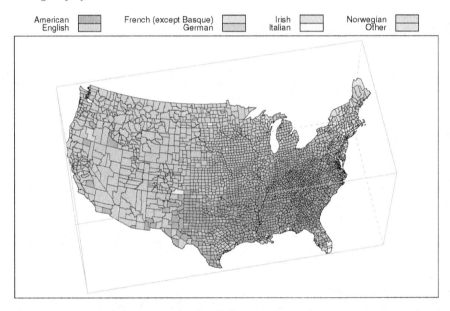

Figure 13.9. Modal ancestry in the U.S. 2000 census by county, using a three-dimensional view to account for the fact that county boundaries lie on a sphere. It is more common to deal with this problem by transforming the boundaries beforehand using one of many cartographic projections.

Average yearly deaths due to cancer per 100000

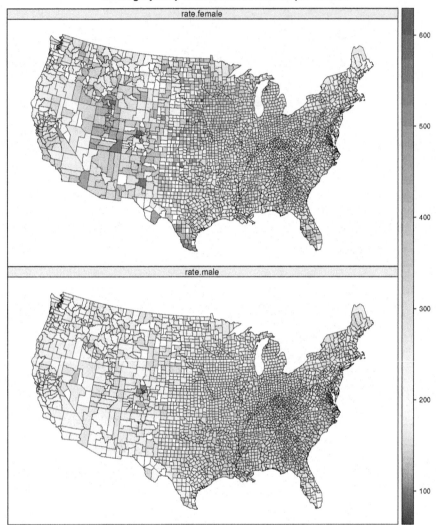

Figure 13.10. Annual death rates due to cancer (1999–2003) in U.S. counties among men and women. A standard projection scheme implemented in the mapproj package is used. The false-color levels are associated with the raw death rates in the color key, but the breakpoints are on a logarithmic scale, resulting in more visual emphasis on variation at the low end.

3.7 Box-and-whisker plots

Two-sample quantile plots can effectively compare two samples at a time, but they do not generalize to more. A matrix of pairwise quantile plots can in principle compare multiple samples, but takes up too much space and can be hard to interpret. A well-known graphical method for comparing multiple samples is the *box-and-whisker plot* (Tukey, 1977). Many variants exist, but essentially each distribution is summarized by five quantiles; three quartiles that define the "box" and two extremes that define the "whiskers". The `bwplot()` function produces box-and-whisker plots with a syntax similar to `qq()`. We illustrate its use by continuing with the `Chem97` example. Figure 3.11 is produced by

```
> bwplot(factor(score) ~ gcsescore | gender, data = Chem97,
        xlab = "Average GCSE Score")
```

Unlike the `qq()` call, the variable defining the samples, `factor(score)` in this case, has more than two levels. For every level, a box-and-whisker plot of the corresponding `gcsescore` values is drawn, allowing us to directly compare the median, indicated by a filled black dot, and the 25*th* and 75*th* quantiles, which determine the range of the box. In some variants, the whiskers extend to the minimum and maximum of the data, but conventionally they are limited to a multiple of the length of the box. This multiple is related to the normal distribution, and points beyond the whiskers, which are plotted explicitly, are thought of as potential outliers; a large number of these indicate tails that are heavier than the normal distribution. These details are controlled by the `coef` and `do.out` arguments of `panel.bwplot()`. In Figure 3.11, the asymmetry in the distribution of `gcsescore` is immediately apparent by looking at the whiskers and the putative outliers, although it is not as clear from the boxes alone.

The next example illustrates the importance of good choices of layout and conditioning. A slightly different version of the previous plot is produced by

```
> bwplot(gcsescore^2.34 ~ gender | factor(score), data = Chem97,
        varwidth = TRUE, layout = c(6, 1),
        ylab = "Transformed GCSE score")
```

The result is shown in Figure 3.12. Although it presents essentially the same data subsets as Figure 3.11 (after applying the optimal Box–Cox transformation to the `gcsescore` values), it orders and orients the boxes differently; in particular, it enables all pairwise comparisons by forcing a common `gcsescore` axis, and emphasizes the differences across `gender` by placing them together. The `varwidth` argument is used to make the widths of the boxes related to sample size, although in this case there is little discernible difference as the sample sizes are all of the same order.

3.7.1 Violin plots

In a sense, the preceding plots summarize all the interesting characteristics of the conditional distribution of `gcsescore`. This usually holds whenever the

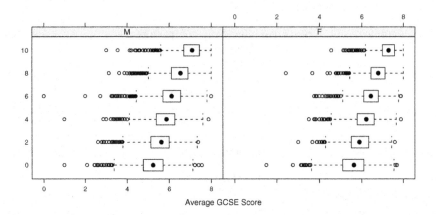

Figure 3.11. Comparative box-and-whisker plots of average GCSE score by final score, conditioned on gender. Systematic skewness and heteroscedasticity, the primary messages of the normal Q–Q plots seen earlier, are readily apparent in this more compact representation as well.

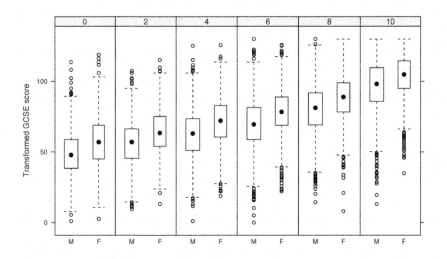

Figure 3.12. Comparative box-and-whisker plots of transformed GCSE scores, representing the same subsets in a slightly different layout. This version highlights a pattern not easily seen in the earlier plots, namely, that boys tend to improve more from bad GCSE scores than girls. This is a good illustration of how layout might affect the information that can be gleaned from a graphic.

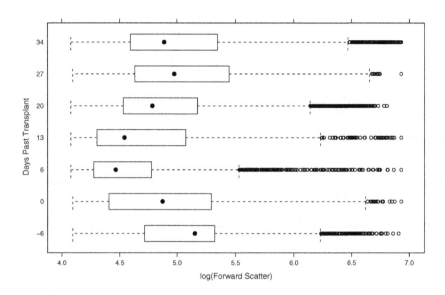

Figure 3.13. Box-and-whisker plots comparing the distribution of log forward scatter values in the **gvhd10** data across time. The multimodality of the distributions, obvious in Figures 3.3 and 3.4, cannot be detected in this encoding.

distribution of interest is unimodal and close to normal. However, box-and-whisker plots can be misleading otherwise. In Figure 3.13, we consider the **gvhd10** data again, this time using a box-and-whisker plot to summarize the distribution of **log(FSC.H)** across **Days**.

```
> bwplot(Days ~ log(FSC.H), data = gvhd10,
         xlab = "log(Forward Scatter)", ylab = "Days Past Transplant")
```

A comparison with Figures 3.3 and 3.4 clearly shows the limitation of this display. A useful alternative that retains the compact structure of a box-and-whisker plot as well as the details of a density plot is the so-called *violin plot* (Hintze and Nelson, 1998). Figure 3.14 is produced by

```
> bwplot(Days ~ log(FSC.H), gvhd10,
         panel = panel.violin, box.ratio = 3,
         xlab = "log(Forward Scatter)",
         ylab = "Days Past Transplant")
```

This uses the predefined panel function **panel.violin()** which can be used as a drop-in replacement for **panel.bwplot()**.

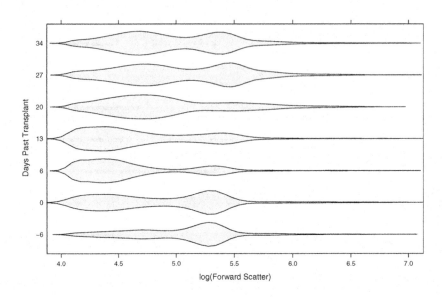

Figure 3.14. A modified version of Figure 3.13, with box-and-whisker plots replaced by violin plots. The bimodal nature of the distributions is readily apparent.

3.8 Strip plots

Box-and-whisker plots summarize the data using a few quantiles, and possibly some outliers. This summarizing can be important when the number of observations is large. When the number of observations per sample is small, it is often sufficient to simply plot the sample values side by side in a common scale. Such plots are known as *strip plots*, also referred to as univariate scatter plots. They are in fact very similar to the bivariate scatter plots we encounter in Chapter 5, except that one of the variables is treated as a categorical variable.

Here, we show a couple of examples using the `quakes` dataset, which records the location (latitude, longitude, and depth) and magnitude of several seismic events near Fiji since 1964. To get a sense of the relationship between magnitude and depth, we might compare the depth values for different magnitudes. Only a few discrete values of magnitude are recorded, and it can be treated as a factor. The call

```
> stripplot(factor(mag) ~ depth, quakes)
```

produces Figure 3.15. There is no particular reason to put the categorical variable on the vertical axis; in fact, the reverse would be the better choice here if we are to have a short wide plot. Figure 3.16 is produced by

```
> stripplot(depth ~ factor(mag), quakes,
            jitter.data = TRUE, alpha = 0.6,
            xlab = "Magnitude (Richter)", ylab = "Depth (km)")
```

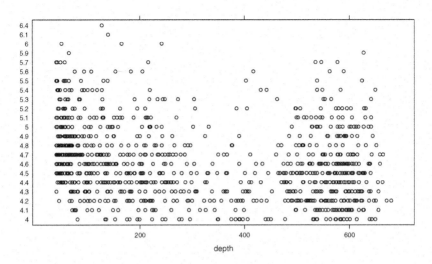

Figure 3.15. Strip plot of depths of the epicenters of seismic events near Fiji, as recorded in the `quakes` dataset. Depths are plotted (on the x-axis) by magnitude of the events on the Richter scale.

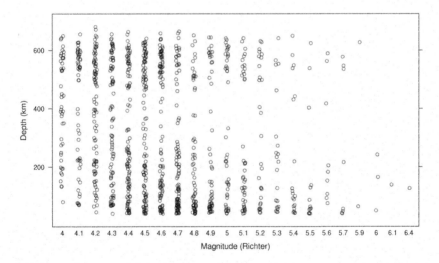

Figure 3.16. Strip plot of epicenter depths by earthquake magnitude, with axes switched. Overplotting is alleviated by jittering the points as well as making them partially transparent.

where additionally we make use of the `alpha` argument to make points semi-transparent[1] and the `jitter.data` argument to randomly displace the points horizontally, both of which help alleviate the effect of overlap.[2] Both plots suggest a weak relationship between depth and magnitude, but the primary visual effect is the clustering of the depth values into two groups, with a gap around 400 km. It is natural to wonder whether this is merely a consequence of some form of spatial clustering of the locations. We follow up on this question in subsequent chapters.

Strip plots can also be used to study residuals from factorial model fits. Figure 3.17, which plots the square roots of the absolute residuals from an additive model fit to the `barley` data, is a variant of the *spread–location* plot (Cleveland, 1993), designed to detect unusual patterns in the variability of residuals.

```
> stripplot(sqrt(abs(residuals(lm(yield~variety+year+site)))) ~ site,
            data = barley, groups = year, jitter.data = TRUE,
            auto.key = list(points = TRUE, lines = TRUE, columns = 2),
            type = c("p", "a"), fun = median,
            ylab = expression(abs("Residual Barley Yield")^{1 / 2}))
```

The call used to produce this plot is somewhat involved, and uses some facts we have not yet encountered. Rather than going into the details now, we wait until we have learned more and analyze the call in Chapter 10.

3.9 Coercion rules

Both the `stripplot()` and `bwplot()` functions expect one of the axes to represent a categorical variable. As with conditioning variables, this can be either a factor or a shingle, and the same coercion rules apply when necessary; that is, a character vector is interpreted as a factor, and a numeric vector as a shingle. The choice of which variable to use as the categorical one is simple when exactly one of the `x` and `y` variables is numeric and the other is a factor or shingle. When the choice is ambiguous, the default is to choose the `y` variable. In all cases, the automatic choice can be overridden by specifying a `horizontal` argument in the high-level call: `TRUE` to have `y` as the categorical variable, `FALSE` to have `x` instead. This choice primarily affects the display produced by the panel function, but also has a subtle effect on axis annotation; by default, for the categorical variable, the axis label is omitted, tick marks are suppressed and the labels do not alternate. These rules also apply to the `dotplot()` and `barchart()` functions discussed in the next chapter.

---

[1] Note that semi-transparency is not supported on all devices.
[2] Both `alpha` and `jitter.data` are actually passed on to the panel function `panel.stripplot()`.

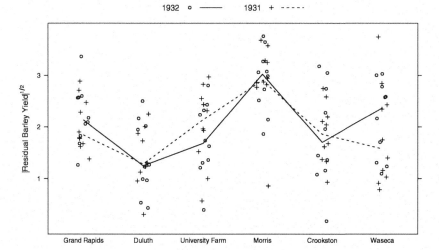

Figure 3.17. A spread–location plot of residuals from a main effects model fit to the `barley` data. The points denote square roots of absolute residuals, jittered horizontally. The lines join the medians of the points within each subgroup, with systematic change in the location indicating a corresponding change in the spread (variance) of the original residuals. It is clear that the model is not entirely appropriate, although the cause is not obvious.

3.10 Discrete distributions

The graphical techniques described in this chapter are designed for continuous random variables. Discrete distributions do not have a density in the conventional sense, but are defined by the analogous probability mass function (p.m.f.). The cumulative distribution function F is still well-defined, as is the quantile function F^{-1} up to certain mathematical caveats. In terms of visualization, this means that density plots and histograms are not really meaningful for discrete data, although Q–Q plots are. The non-parametric maximum likelihood estimator of a p.m.f. is the (relative) frequency table. As with other tables, these can be visualized using bar charts and dot plots (Chapter 4), which serve as a substitute for density plots and histograms; in fact, bar charts are often loosely referred to as histograms, although we prefer not to confuse the two. As with its continuous analogues, bar charts are not very effective tools for judging the goodness of fit of a reference distribution. An innovative visualization for that purpose is Tukey's hanging rootogram (Cleveland, 1988); we do not encounter them in this book, but Trellis rootograms can be created by the `rootogram()` function in the latticeExtra package.

The distinction between continuous and discrete distributions is sometimes unclear. The `gcsescore` variable in the `Chem97` serves as a good case in point;

the 31,022 values, ranging from 0 to 8 with a resolution of three digits after the decimal point, have only 244 unique values. Figure 4.8 in the next chapter, which treats the variable as discrete, reveals an apparent rounding artifact; this is also noticeable in some of the displays we have already seen, but only if we know what we are looking for. Note in particular the ECDF plot in Figure 3.8, which is simply a cumulative version of the bar chart in Figure 4.8, up to differences in conditioning. This emphasizes the important point that two graphical encodings with the same "information" can put visual emphasis on entirely different aspects of that information.

Some distributions are neither continuous nor discrete, but a mixture of the two. These are mostly irrelevant in practical data analysis, with the important exception of *censoring*. Formally, censoring refers to the situation where only the range (most commonly an upper or lower bound) of an observation is known and not its exact value. Often, the fact that an observation is censored is also known, and this can be taken into account during analysis. In other cases, however, censored values may be silently encoded as the bound (which may be the limit of a measuring instrument, for instance), leading to a discrete point mass in an otherwise continuous distribution. Such situations are often hard to identify graphically with density plots or bar charts; the best bet is the Q–Q plot as F and F^{-1} are still well-defined even though the density and p.m.f. are not. Censoring effects can be seen in Figure 10.4.

3.11 A note on the formula interface

We end this chapter with a remark on the formula interface. Although the lattice interface is similar in many ways to the one used in statistical modeling functions in S and R, the interpretation of terms in the formula differs substantially; in fact, the interpretation is not even consistent across lattice functions. A generally helpful rule is the following: given a formula such as y ˜ x, the y variable will be plotted on the y-axis and the x variable on the x-axis. The exception to this rule is qq(). Similarly, most functions with a formula of the form ˜ x plot x on the x-axis, with the exception of qqmath() and the yet to be seen splom() and parallel(). Formulae in the trivariate functions described in Chapter 6 have the form z ˜ x * y, where again a similar association holds with certain caveats. The upshot is that there is no single rule that governs all uses; the formula interface should be simply viewed as a convenient language that defines the structure of a lattice graphic, and is to be interpreted only in the context of that particular graphic.

4
Displaying Multiway Tables

An important subset of statistical data comes in the form of tables. Tables usually record the frequency or proportion of observations that fall into a particular category or combination of categories. They could also encode some other summary measure such as a rate (of binary events) or mean (of a continuous variable). In R, tables are usually represented by arrays of one (vectors), two (matrices), or more dimensions. To distinguish them from other vectors and arrays, they often have class *"table"*. The R functions `table()` and `xtabs()` can be used to create tables from raw data.

Graphs of tables do not always convey information more easily than the tables themselves, but they often do. The `barchart()` and `dotplot()` functions in lattice are designed to display tabulated data. As with other high-level functions, the primary formula interface requires the data to be available as a data frame. The `as.data.frame.table()` function can be used for converting tables to suitable data frames. In addition, there are methods in lattice that work directly on tables. We focus on the latter in this chapter; examples using the formula interface can be found in Chapter 2.

4.1 Cleveland dot plot

Dot plots (Cleveland, 1985) provide simple and effective graphical summaries of tables that are perhaps less often used than they should be. For illustration, we use the `VADeaths` data, which is a cross-classification of death rates in the U.S. state of Virginia in 1940 by age and population groups (Molyneaux et al., 1947).

```
> VADeaths
      Rural Male Rural Female Urban Male Urban Female
50-54       11.7          8.7       15.4          8.4
55-59       18.1         11.7       24.3         13.6
60-64       26.9         20.3       37.0         19.3
```

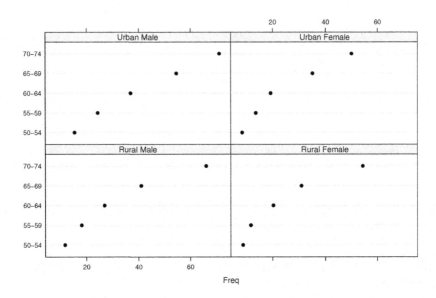

Figure 4.1. Dot plots of death rates (per 1000) in Virginia in 1940, cross-tabulated by age and demographic groups.

| 65–69 | 41.0 | 30.9 | 54.6 | 35.1 |
| 70–74 | 66.0 | 54.3 | 71.1 | 50.0 |

The **VADeaths** object is of class *"matrix"*.

```
> class(VADeaths)
[1] "matrix"
```

We can check what methods are available for the **dotplot()** function using

```
> methods("dotplot")
[1] dotplot.array*   dotplot.default* dotplot.formula*
[4] dotplot.matrix*  dotplot.numeric* dotplot.table*
```

   Non-visible functions are asterisked

As we can see, there is a method for *"matrix"* objects, which we can use directly in this case. The corresponding help page can be viewed by typing **help(dotplot.matrix)**. Figure 4.1 is produced by

```
> dotplot(VADeaths, groups = FALSE)
```

which uses one additional argument to disable grouping. As is almost inevitable with a first attempt, there is much scope for improvement. The default label on the horizontal axis says **Freq**, even though the table values are not frequencies. More importantly, this display does not easily allow us to compare the rates for males and females, as they are displayed in different columns. One way to rectify this is to force the display to have one column. To prevent the panels from getting too flattened, we add an explicit aspect ratio. Because rates have a well-defined origin (0), it may also be interesting to make a judgment about their relative magnitude, and not just their differences. To this end, we ask for the points to be joined to a baseline using the `type` argument. Figure 4.2 is produced by

```
> dotplot(VADeaths, groups = FALSE,
          layout = c(1, 4), aspect = 0.7,
          origin = 0, type = c("p", "h"),
          main = "Death Rates in Virginia - 1940",
          xlab = "Rate (per 1000)")
```

Even more direct comparison can be achieved using superposition, which is in fact the default in this `dotplot()` method. By omitting the `groups = FALSE` argument, we can plot rates for all the population groups in a single panel, but with different graphical parameters. The following call produces Figure 4.3.

```
> dotplot(VADeaths, type = "o",
          auto.key = list(lines = TRUE, space = "right"),
          main = "Death Rates in Virginia - 1940",
          xlab = "Rate (per 1000)")
```

4.2 Bar chart

Bar charts (along with pie charts[1]) are among the most popular graphical representations of tables. However, they are less useful than dot plots in most situations. A bar chart analogous to the dot plot in Figure 4.1 is produced by

```
> barchart(VADeaths, groups = FALSE,
           layout = c(1, 4), aspect = 0.7, reference = FALSE,
           main = "Death Rates in Virginia - 1940",
           xlab = "Rate (per 100)")
```

The resulting plot, shown in Figure 4.4, conveys exactly the same information with some additional and redundant graphical structure. In fact, bar charts can actually mislead when the "origin" is arbitrary, as they convey the incorrect impression that the quantity encoded by the length (or area) of the bar has some meaning. Another popular but questionable practice is to add confidence intervals to bar charts; dot plots with confidence intervals are almost invariably easier to interpret.

One variant of the bar chart does encode more information than a dot plot could. A grouping variable can be incorporated in a bar chart display either by plotting the bars for the various groups side by side or by stacking

---

[1] lattice does not contain a function that produces pie charts. This is entirely by choice, as pie charts are a highly undesirable form of graphical representation (see Cleveland (1985) for a discussion), and their use is strongly discouraged.

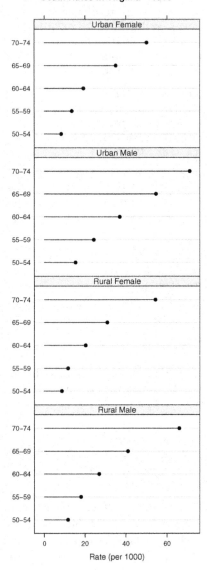

Figure 4.2. Dot plot of death rates in Virginia in 1940, arranged in a single column layout with more informative labels. The origin is included in the plot, and points are joined to it to enable comparison of absolute rates.

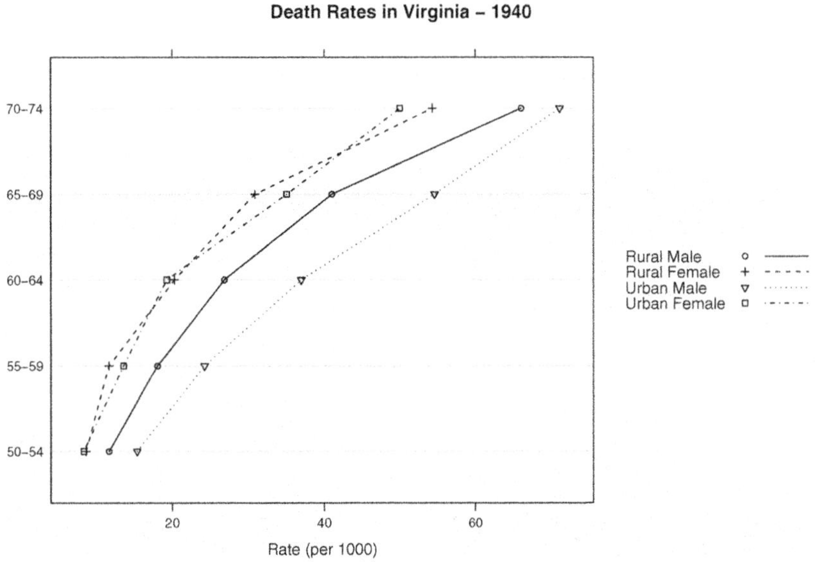

Death Rates in Virginia – 1940

Figure 4.3. Death rates in Virginia in 1940, with population groups superposed within a single panel. Points within a group are joined to emphasize group membership. This plot suggests that the rates are virtually identical in the rural female and urban female subgroups, with a systematic increase among rural males and a further increase for urban males. This pattern is hard to see in the multipanel versions.

them on top of each other. The first case is similar to a grouped dot plot and contains no extra information. In the second case, a stacked bar chart, the total length of each bar encodes the marginal totals, in addition to the lengths of the component bars, which breaks up this total according to the grouping variable. We have seen stacked bar charts previously in Chapter 2 (e.g., Figure 2.9). For another example, consider the data in Table 4.2, based on a survey of doctorate degree recipients in the United States who went on to pursue a postdoctoral position. The data are available in the latticeExtra package.

```
> data(postdoc, package = "latticeExtra")
```

Stacked bar charts are generally produced by adding a `stack` = TRUE argument to `barchart()`, but this is unnecessary for the *"table"* method as it is the default. A stacked bar chart of the `postdoc` data, shown in Figure 4.5, is produced by

```
> barchart(prop.table(postdoc, margin = 1), xlab = "Proportion",
          auto.key = list(adj = 1))
```

The data plotted are proportions, computed by `prop.table()`, as these are the quantities of interest; the counts could have been plotted as well, but that would not have told us much except that the "Biological Sciences" field

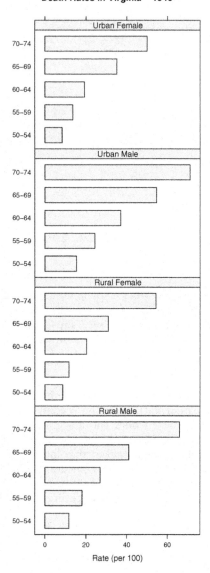

Figure 4.4. Bar charts of Virginia death rates by population group, in a layout similar to the dot plots in Figure 4.1. This encoding contains more graphical structure, but no more information.

	Expected or Additional Training	Work with Specific Person	Training Outside PhD Field	Other Employment Not Available	Other
Biological Sciences	6404	2427	1950	1779	602
Chemistry	865	308	292	551	168
Earth, Atm., & Ocean Sciences	343	75	75	238	80
Engineering	586	464	288	517	401
Medical Sciences	205	137	82	68	74
Physics & Astronomy	1010	347	175	399	162
Social & Behavioral Sciences	1368	564	412	514	305
All Postdoctorates	11197	4687	3403	4406	1914

Table 4.1. Reasons for choosing a postdoctoral position after graduating from U.S. universities, by different fields of study.

contributes the majority of postdocs. We also make the levels of the grouping variable right-justified in the legend using the `auto.key` argument.

A multipanel dot plot encoding the same information can be produced by

```
> dotplot(prop.table(postdoc, margin = 1), groups = FALSE,
          xlab = "Proportion",
          par.strip.text = list(abbreviate = TRUE, minlength = 10))
```

creating Figure 4.6. Even though the stacked bar chart is more concise, it is not necessarily better if one is primarily interested in comparing the proportions of reasons across fields. The bar chart encodes this quantity using length, whereas the dot plot does so using relative position which is more easily judged by the human eye.

4.2.1 Manipulating order

A point worth making in the context of this example is the importance of visual order. In many situations, the levels of a categorical variable have no natural order; this is true for both margins of the `postdoc` table. Often, choosing the order in which these levels are displayed based on the data can significantly increase the impact of the display. An effective order is usually obtained by sorting levels by the value of a corresponding continuous response, or perhaps a summary measure when multiple observations or multiple responses are involved. Facilities available in lattice that aid such reordering are discussed, along with examples, in Chapter 10. Unfortunately, our example is slightly complicated by the fact that the responses are proportions that add up to one for each field of study, making it difficult to find a common order that is appropriate for all panels.

One solution is to use a different order for each panel. This is not particularly difficult to achieve, but involves several concepts we have not yet

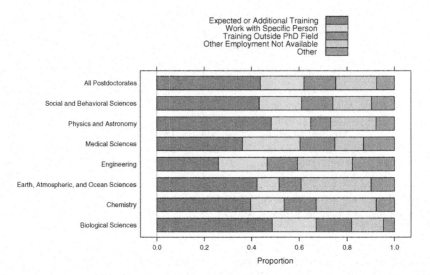

Figure 4.5. A stacked bar chart showing the proportion of reasons for choosing a postdoc by field of study. Because the bars encode proportions, their lengths add up to one within each field. Comparison is done through lengths (except for the first and last group), which is less effective than comparison through position. Notice the long labels on the vertical axis, for which enough space has been allocated automatically.

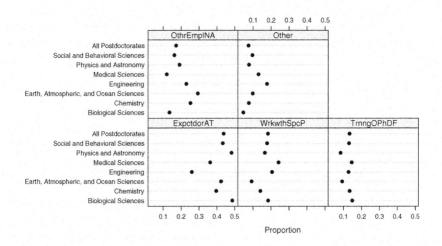

Figure 4.6. Reasons for choosing a postdoc position; an alternative visualization using multipanel dot plots. Although the display is less compact, it makes comparison within fields easier. Long axis labels are not uncommon in situations such as these, therefore the labels are all shown on one side by default to save space. The strip labels have been abbreviated as they would not have fit in the available area.

encountered. We give the solution here for the sake of completeness, and refer the reader to later chapters for details. Figure 4.7 is produced by

```
> dotplot(prop.table(postdoc, margin = 1), groups = FALSE,
          index.cond = function(x, y) median(x),
          xlab = "Proportion", layout = c(1, 5), aspect = 0.6,
          scales = list(y = list(relation = "free", rot = 0)),
          prepanel = function(x, y) {
              list(ylim = levels(reorder(y, x)))
          },
          panel = function(x, y, ...) {
              panel.dotplot(x, reorder(y, x), ...)
          })
```

The critical additions in this call are the use of the `index.cond` argument and the `reorder()` function, both of which are discussed in Chapter 10. The order of the fields is changed inside the panel function, and a corresponding change is required in the `prepanel` function to ensure that the axis labels match. We also need to specify an appropriate `scales` argument to allow panels to have independent axis annotation. Details about these arguments can be found in Chapter 8.

4.2.2 Bar charts and discrete distributions

As mentioned in Chapter 3, bar charts can be viewed as analogues of density plots or histograms for discrete distributions. In the examples we have seen so far, the data come in the form of a table. When only raw data are available, frequency tables can be easily constructed with the `xtabs()` function. Consider the following two-way table of `gcsescore` by `gender` derived from the `Chem97` data.

```
> gcsescore.tab <- xtabs(~gcsescore + gender, Chem97)
```

We might attempt to produce a bar chart directly from this table, but this will not give us the result we want; because `gcsescore` is now interpreted as a categorical variable, its levels will be plotted as equispaced integers[2] and not as the original numeric values. Additionally, `barchart()` will print the labels for every level of `gcsescore`, causing substantial overlap. In this case, it is easier to first manipulate the data, after converting the table into a data frame, and then use the `xyplot()` function, which does not require either of the variables to be categorical. The next chapter discusses `xyplot()` in detail; here we use the convenient `type` argument to create Figure 4.8.

```
> gcsescore.df <- as.data.frame(gcsescore.tab)
> gcsescore.df$gcsescore <-
      as.numeric(as.character(gcsescore.df$gcsescore))
> xyplot(Freq ~ gcsescore | gender, data = gcsescore.df,
         type = "h", layout = c(1, 2), xlab = "Average GCSE Score")
```

---

[2] To be precise, the numeric codes in the underlying representation of a *"factor"*.

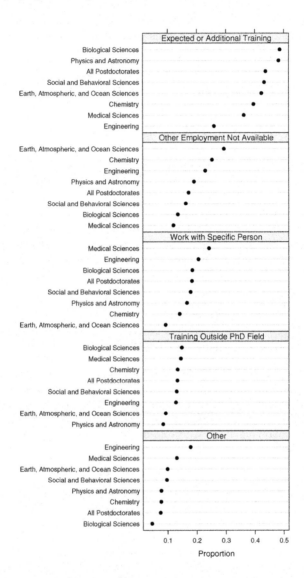

Figure 4.7. Yet another visualization of reasons for choosing a postdoc. Both margins have been ordered by the response (proportions within field): the panels (reasons) are ordered by the median proportion over all fields, and fields are ordered by proportion within each panel. Reordering often makes it easier to see patterns in the data when there is no intrinsic order.

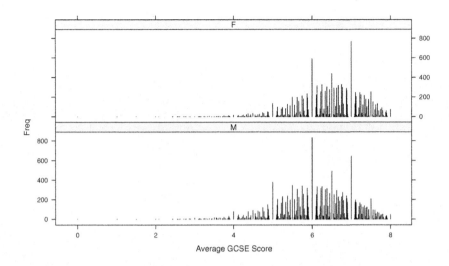

Figure 4.8. A bar chart of sorts, visualizing the frequency table of average GCSE score by gender. The main differences from a conventional bar chart are that the x-axis is continuous and the "bars" are actually zero-width lines. Apart from showing that girls tend to do better than boys on the GCSE, the most interesting feature of the display is the spikes of high frequencies for certain values, most noticeable for whole numbers. At first glance, this might appear to be some sort of rounding error. In fact, the artifact is due to averaging; most of the GCSE score values are the average of 8, 9, or 10 scores, where the total scores are integers.

Note that the use of `barchart()` is perfectly reasonable when the number of levels is small; for example, Figure 4.9 is produced by

```
> score.tab <- xtabs(~score + gender, Chem97)
> score.df <- as.data.frame(score.tab)
> barchart(Freq ~ score | gender, score.df, origin = 0)
```

4.3 Visualizing categorical data

Tables are examples of the more general class of categorical data. Specialized visualization methods for such data exist, but are less well known compared to methods for continuous data. Support for visualizing categorical data in lattice is limited to dot plots and bar charts, and those interested in such data are strongly encouraged to look at the vcd package, which implements many techniques described by Friendly (2000).

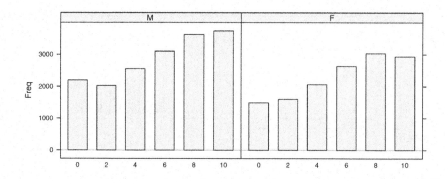

Figure 4.9. Bar chart displaying the frequency distribution of final score in the A-level chemistry examination, by gender.

5
Scatter Plots and Extensions

The scatter plot is possibly the single most important statistical graphic. In this chapter we discuss the xyplot() function, which can be used to produce several variants of scatter plots, and splom(), which produces scatter-plot matrices. We also include a brief discussion of parallel coordinates plots, as produced by parallel(), which are related to scatter-plot matrices in terms of the kinds of data they are used to visualize, although not so much in the actual visual encoding.

A *scatter plot* graphs two variables directly against each other in a Cartesian coordinate system. It is a simple graphic in the sense that the data are directly encoded without being summarized in any way; often the aspects that the user needs to worry about most are graphical ones such as whether to join the points by a line, what colors to use, and so on. Depending on the purpose, scatter plots can also be enhanced in several ways. In this chapter, we go over some of the variants supported by panel.xyplot(), which is the default panel function for both xyplot() and splom() (under the alias panel.splom()).

5.1 The standard scatter plot

We continue with the quakes example from Chapter 3. We saw in Figure 3.16 that the depths of the epicenters more or less fall into two clusters. The latitude and longitude are also recorded for each event, and together with depth could provide a three-dimensional view of how the epicenters are spatially distributed. Of course, scatter plots can only show us two dimensions at a time. As a first attempt, we could divide up the events into two groups by depth and plot latitude against longitude for each. Figure 5.1 is created using the by now familiar formula interface.

```
> xyplot(lat ~ long | cut(depth, 2), data = quakes)
```

The cut() function is used here to convert depth into a discrete factor by dividing its range into two equal parts. There does indeed seem to be some

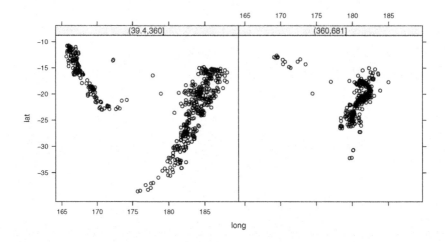

Figure 5.1. Scatter plots of latitude against longitude of earthquake epicenters, conditioned on depth discretized into two groups. The distribution of locations in the latitude–longitude space is clearly different in the two panels.

differentiation in the two parts; the cluster of locations towards the upper-left corner all but disappears in the second panel. The other cluster also appears to shrink, but it is not immediately clear if there is a spatial shift as well.

For our second attempt, we make several changes. We discretize the `depth` values into three groups instead of two, hoping to discern some finer patterns. We use a variant of the default strip function so that the name of the conditioning variable is included in the strips. We change the plotting symbol to dots rather than circles. Because the two axes have the same units (degrees), we constrain the scales to be isometric by specifying `aspect = "iso"`, which forces the aspect ratio to be such that the relationship between the physical and native coordinate systems (the number of data units per cm) is the same on both axes (of course, this does not account for the locations falling on a sphere and not a plane). Finally, and perhaps most important, we add a common reference grid to all three panels. Figure 5.2 is produced by

```
> xyplot(lat ~ long | cut(depth, 3), data = quakes,
         aspect = "iso", pch = ".", cex = 2, type = c("p", "g"),
         xlab = "Longitude", ylab = "Latitude",
         strip = strip.custom(strip.names = TRUE, var.name = "Depth"))
```

Thanks to the reference grid, careful inspection now confirms a subtle but systematic spatial pattern; for example, consider the neighbourhood of the $(185, -20)$ grid location in the three panels. Grids and other common (not data driven) reference objects are often invaluable in multipanel displays.

As we have seen in other contexts, superposition offers more direct between group comparison when it is feasible. In Figure 5.3 we show a grouped display with a slight variation; we discretize `depth` into three groups as before, but

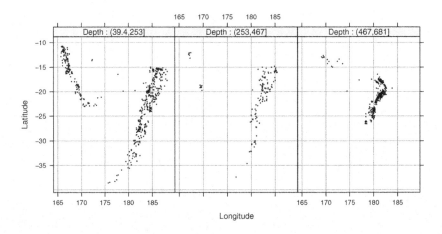

Figure 5.2. A slight variant of the previous plot. Depth is now discretized into three groups, a smaller plotting character reduces overlap, and a reference grid makes it easier to see trends across panels. In addition, the aspect ratio is such that the scales are now "isometric" (i.e., the number of data units per cm is the same on both axes). This aspect ratio is retained even when an on screen rendering is resized.

use equispaced quantiles as breakpoints, ensuring that all three groups have roughly the same number of points.

```
> xyplot(lat ~ long, data = quakes, aspect = "iso",
         groups = cut(depth, breaks = quantile(depth, ppoints(4, 1))),
         auto.key = list(columns = 3, title = "Depth"),
         xlab = "Longitude", ylab = "Latitude")
```

Although these examples all consistently hint at a certain spatial pattern, they all discretize the continuous **depth** variable. An obvious extension to this idea is to encode **depth** by a continuous gradient of some sort; color and symbol size are the most common choices. The human eye does not make very good quantitative judgments from such encodings, but relative ordering is conveyed reasonably well. In Figure 5.4, we use shades of grey to encode depth. There is no built-in support to achieve this in **xyplot()**, and we need to first create a suitable vector of colors to go with each observation. To this end, we use **cut()** again to convert **depth** into an integer code that is used to index a vector of colors.

```
> depth.col <- grey.colors(100)[cut(quakes$depth, 100, label = FALSE)]
```

We also reorder the rows to ensure that shallower points are plotted after deeper points; this requires us to reorder the color vector as well to keep the association between rows and colors valid. Figure 5.4 is produced by

```
> depth.ord <- rev(order(quakes$depth))
> xyplot(lat ~ long, data = quakes[depth.ord, ],
```

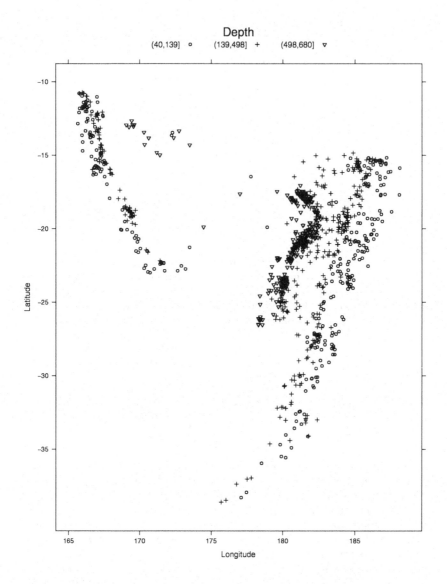

Figure 5.3. Scatter plots of latitude against longitude of earthquake epicenters. Depth, discretized into three slightly different groups, is now indicated using different plotting symbols within a single panel.

```
aspect = "iso", type = c("p", "g"), col = "black",
pch = 21, fill = depth.col[depth.ord], cex = 2,
xlab = "Longitude", ylab = "Latitude")
```

This simple approach works in this case; however, it does not generalize to multipanel displays. Attempting to add a conditioning variable will lead to the same color vector being used in each panel, thus losing the correspondence between colors and rows in `quakes`.

5.2 Advanced indexing using `subscripts`

Fortunately, this is a common enough situation that a standard solution exists. It does, however, require the use of a simple panel function, and the reader is encouraged to revisit Section 2.5.3 before proceeding.

Our goal in this section is to create a multipanel version of Figure 5.4. A natural choice for a conditioning variable is `mag`, which gives the magnitude of each earthquake on the Richter scale, as we may be interested in knowing if the location of a quake has any relation to its magnitude. As `mag` is a continuous variable, we need to discretize it, just as we did with `depth`. However, this time, instead of `cut()`, we use `equal.count()` to create a *shingle*.

```
> quakes$Magnitude <- equal.count(quakes$mag, 4)
> summary(quakes$Magnitude)

Intervals:
   min  max count
1 3.95 4.55   484
2 4.25 4.75   492
3 4.45 4.95   425
4 4.65 6.45   415


Overlap between adjacent intervals:
[1] 293 306 217
```

As mentioned in Chapter 2, shingles are generalizations of factors for continuous variables, with possibly overlapping levels, allowing a particular observation to belong to more than one level. The `equal.count()` function creates shingles with overlapping levels that each have roughly the same number of observations (hence the name `equal.count`). The newly created `Magnitude` variable can now be used as a conditioning variable. By default, the intervals defining levels of a shingle relative to its full range are indicated by a shaded rectangle in the strip. To produce Figure 5.5, we use a call similar to the last one (this time creating a data frame with the desired row order beforehand), but with an explicit panel function.

```
> quakes$color <- depth.col
> quakes.ordered <- quakes[depth.ord, ]
> xyplot(lat ~ long | Magnitude, data = quakes.ordered, col = "black",
         aspect = "iso", fill.color = quakes.ordered$color, cex = 2,
```

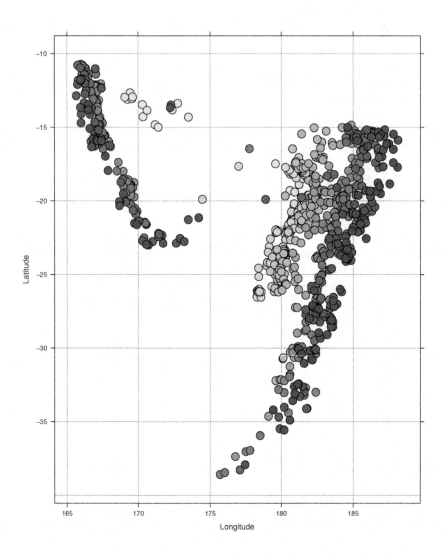

Figure 5.4. Latitude and longitude of earthquake epicenters, with the continuous depth variable encoded by fill color. A legend that describes the association between grey levels and the depths they represent would be a useful addition, but this is slightly more difficult. We show an example of such a legend in Figure 5.6.

```
panel = function(x, y, fill.color, ..., subscripts) {
    fill <- fill.color[subscripts]
    panel.grid(h = -1, v = -1)
    panel.xyplot(x, y, pch = 21, fill = fill, ...)
},
xlab = "Longitude", ylab = "Latitude")
```

Before looking at the panel function, note the argument fill.color which contains the vector of colors corresponding to rows of the full data frame. As explained in Section 2.5.3, xyplot() will pass this argument on to the panel function as it does not recognize it itself. Thus, every time the panel function gets executed, it has access to the full vector of colors.

The problem of course is that the x and y values in the panel function only represent the subset of rows in that panel and not the full data. To use the colors correctly, we need to extract the colors associated with this subset from the full color vector fill.color. This is where the subscripts argument comes in. Along with other arguments, xyplot() can provide the panel function with an argument called subscripts containing a vector of integer indices that give the row numbers of the corresponding primary variables (x and y in this case). In other words, the correct color vector to go with x and y in a panel is fill.color[subscripts]; this fact is used in the panel function above to obtain the correct colors.

While we are discussing subscripts, we should note that the groups argument, already used in many examples, is essentially no different from the fill.colors argument used above; it simply gets passed on to the panel function in its entirety. The only thing special about groups, other than the fact that certain panel functions treat it specially, is that it gets evaluated in data. This is not true for other arguments, which is why we had to specify fill.colors explicitly as quakes.ordered$color. In fact, we can simply replace all references to fill.color by groups and obtain the same results, as in the following call that produces Figure 5.6,[1] with a slightly different color calculation that uses the convenient level.colors() function.

```
> depth.breaks <- do.breaks(range(quakes.ordered$depth), 50)
> quakes.ordered$color <-
      level.colors(quakes.ordered$depth, at = depth.breaks,
                   col.regions = grey.colors)
> xyplot(lat ~ long | Magnitude, data = quakes.ordered,
         aspect = "iso", groups = color, cex = 2, col = "black",
         panel = function(x, y, groups, ..., subscripts) {
             fill <- groups[subscripts]
             panel.grid(h = -1, v = -1)
             panel.xyplot(x, y, pch = 21, fill = fill, ...)
         },
         legend =
```

---

[1] Of course, the name groups is misleading in this example, and in any case this will not work when there are two or more variables to pass. See Figure 9.2 for such an example.

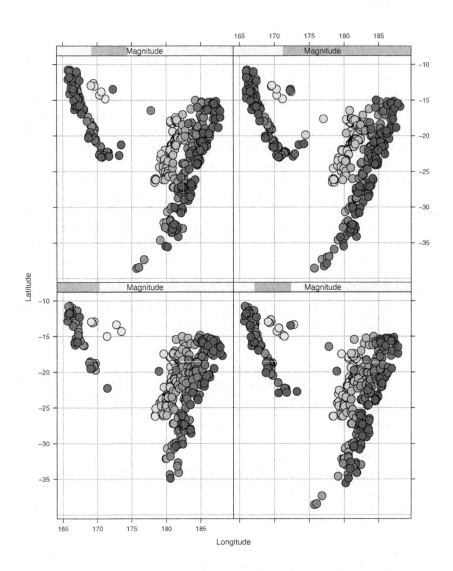

Figure 5.5. A multipanel version of Figure 5.4, conditioning on overlapping subsets of magnitudes.

type	Effect	Panel function
"p"	Plot points	
"l"	Join points by lines	
"b"	Both points and lines	
"o"	Points and lines overlaid	
"S", "s"	Plot as step function	
"h"	Drop lines to origin ("histogram-like")	
"a"	Join by lines after averaging	panel.average()
"r"	Plot regression line	panel.lmline()
"smooth"	Plot LOESS smooth	panel.loess()
"g"	Plot a reference grid	panel.grid()

Table 5.1. The effect of various values of the `type` argument in `panel.xyplot()`. For some values, the effect will also depend on the value of the `horizontal` argument, as seen in Figure 5.7. Effects can be (and usually are) combined by specifying `type` as a vector. The actual rendering for some of these effects is performed by other specialized panel functions, and having access to them through the `type` argument is simply a convenience. The `type` argument also works for grouped displays transparently; when the `groups` argument is specified, `panel.xyplot` automatically calls another specialized panel function, `panel.superpose()`, to handle the necessary details.

```
list(right =
     list(fun = draw.colorkey,
          args = list(key = list(col = grey.colors,
                                 at = depth.breaks),
                      draw = FALSE))),
xlab = "Longitude", ylab = "Latitude")
```

Here, to make things interesting, we have also added a color key linking the colors to the depth values. This is somewhat nontrivial because `xyplot()` does not explicitly support such legends. What we have done, in fact, is to use a very general feature of lattice where an arbitrary legend can be specified in terms of a function that creates it. In this case, the relevant function is `draw.colorkey()`, which is called with arguments `key` and `draw` as specified in the call above. To learn more about this feature, consult Chapter 9 and the online documentation.

5.3 Variants using the `type` argument

As we have already seen, the `type` argument can be used to add a reference grid to each panel. It can also be used for a variety of other enhancements. Although it is typically supplied directly to `xyplot()`, it is actually an argument of the default panel function `panel.xyplot()`. Valid values of `type` and their effects are summarized in Table 5.1 and Figure 5.7. Its most common use is as `type = "l"` to plot lines instead of points (e.g., for time-series data). It is often supplied as a vector, in which case the effects of the individual components

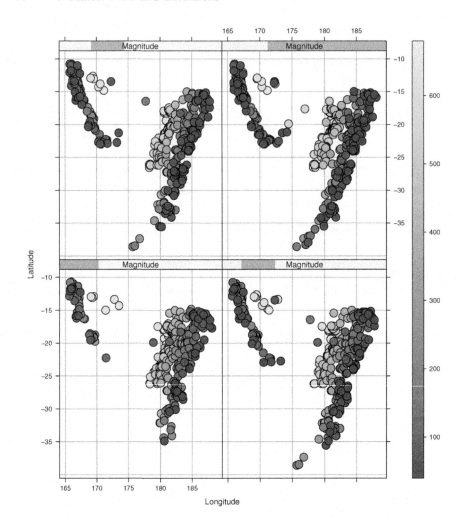

Figure 5.6. Variant of Figure 5.5, with a key describing the encoding of depth by fill color.

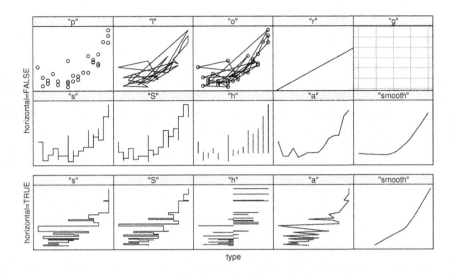

Figure 5.7. The effect of various values of **type** when specified as an argument to xyplot(), as well as dotplot(), stripplot(), and splom(). In each of these cases, **type** is eventually passed on to **panel.xyplot()** which does the actual plotting. Some of the types (e.g., "s", "S", and "a") sort the data first. The step types "s" and "S" differ from each other by whether the first move is vertical or horizontal. The behavior for some types depends on the value of **horizontal**; this is more relevant for dotplot() and stripplot() where **horizontal** is set to TRUE automatically when the y variable is a factor. An example can be seen in Figure 4.2. The "a" type can be useful in creating interaction plots in conjunction with a **groups** argument.

are combined (except in certain grouped displays; see Figure 5.12). Some of the values (e.g., "r", "g", and "smooth") simply cause other predefined panel functions to be called, and are provided as a convenience. As an example, consider another dataset on earthquakes, this one available in the MEMSS package, consisting of seismometer measurements of 23 large earthquakes in North America (Joyner and Boore, 1981).

```
> data(Earthquake, package = "MEMSS")
```

Ignoring the fact that multiple measurements are recorded from each earthquake, we wish to explore how the maximum horizontal acceleration at a measuring center (accel) depends on its distance from the epicenter (distance). It is fairly common to include a reference grid and a LOESS smooth (Cleveland and Devlin, 1988; Cleveland and Grosse, 1991) in such scatter plots. Without using the **type** argument, we could call

```
> xyplot(accel ~ distance, data = Earthquake,
         panel = function(...) {
             panel.grid(h = -1, v = -1)
             panel.xyplot(...)
```

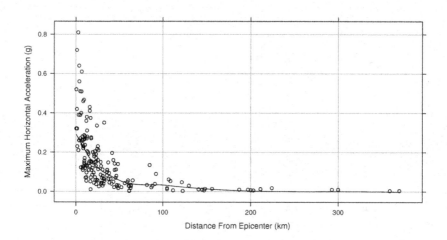

Figure 5.8. Scatter plot of acceleration versus distance in the `Earthquake` data, with a reference grid and a LOESS smooth. The asymmetry in the distribution of points on both axes, with only a few large values, suggests that a transformation is required.

```
        panel.loess(...)
},
xlab = "Distance From Epicenter (km)",
ylab = "Maximum Horizontal Acceleration (g)")
```

This produces Figure 5.8. It is clear that transforming the data should improve the plot, and because both quantities are positive, we try plotting them on a logarithmic scale next in Figure 5.9. This time, however, we use the `type` argument instead of a custom panel function to get the equivalent result.

```
> xyplot(accel ~ distance, data = Earthquake,
         type = c("g", "p", "smooth"),
         scales = list(log = 2),
         xlab = "Distance From Epicenter (km)",
         ylab = "Maximum Horizontal Acceleration (g)")
```

This approach allows for concise and more readable code. It also avoids the concept of a panel function, which can be daunting for R beginners, while exposing some of its power. Of course, the disadvantage is that one is limited to the functionality built in to `panel.xyplot()`. Figure 5.10, produced by the following call, splits the data into three panels depending on the magnitude of the quakes, adds a common reference regression line to each panel, and uses an alternative smoothing method from the locfit package (Loader, 1999).

```
> library("locfit")
> Earthquake$Magnitude <-
      equal.count(Earthquake$Richter, 3, overlap = 0.1)
```

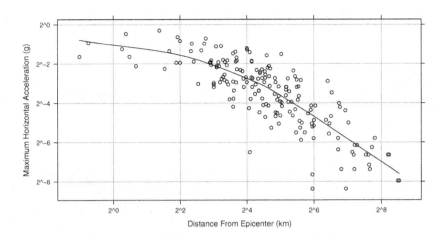

Figure 5.9. Scatter plot of acceleration versus distance on a logarithmic scale. The relationship between the variables is much more obvious in this plot. The axis labeling could be improved; this issue is taken up in Chapter 8.

```
> coef <- coef(lm(log2(accel) ~ log2(distance), data = Earthquake))
> xyplot(accel ~ distance | Magnitude, data = Earthquake,
         scales = list(log = 2), col.line = "grey", lwd = 2,
         panel = function(...) {
             panel.abline(reg = coef)
             panel.locfit(...)
         },
         xlab = "Distance From Epicenter (km)",
         ylab = "Maximum Horizontal Acceleration (g)")
```

This simple yet useful plot would not have been possible without a custom panel function.

5.3.1 Superposition and `type`

The `type` argument is useful in grouped displays as well. By default, it is interpreted just as described earlier; each component of `type` is used for each level of `groups`, with different graphical parameters. However, this is not always the desired behavior. Consider the `SeatacWeather` dataset in the latticeExtra package, which records daily temperature and rainfall amounts at the Seattle–Tacoma airport in the U.S. state of Washington over the first three months of 2007.

```
> data(SeatacWeather, package = "latticeExtra")
```

Suppose that we wish to plot the daily minimum and maximum temperatures as well as the daily rainfall in a single plot, with one panel for each month.

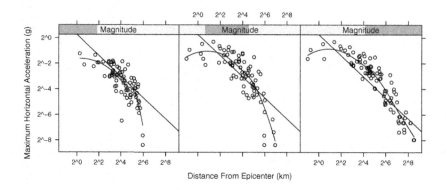

Figure 5.10. Scatter plots of acceleration by distance conditioned on earthquake magnitude. The common reference line makes it easier to see the shift across panels. The smooths are computed and plotted by the `panel.locfit()` function in the locfit package.

Getting all the variables into a single panel is simple if we use the extended formula interface described in Chapter 10; Figure 5.11 is produced by

```
> xyplot(min.temp + max.temp + precip ~ day | month,
         ylab = "Temperature and Rainfall",
         data = SeatacWeather, type = "l", lty = 1, col = "black")
```

The `lty` and `col` arguments are explicitly specified to prevent `panel.xyplot()` from using different ones for the three groups, which does not really help in this example. There are two problems with this plot. First, the rainfall measurements are in a completely different scale. Second, even though this is not obvious from Figure 5.11, most of the rainfall measurements are 0, which is special in this context, and joining the daily rainfall values by lines does not reflect this point. The first problem can only be solved by rescaling the rainfall values for the purpose of plotting (this brings up the issue of axis labeling, which we deal with later). As for the second problem, `type = "h"` seems to be the right solution. Thus, we would like to use `type = "l"` as before for the first two groups (`min.temp` and `max.temp`), and `type = "h"` for the third (`precip`). This can be achieved using the `distribute.type` argument[2] which, when TRUE, changes the interpretation of `type` by using the first component for the first level of `groups`, the second component for the second level, and so on. Figure 5.12 is produced by

```
> maxp <- max(SeatacWeather$precip, na.rm = TRUE)
> xyplot(min.temp + max.temp + I(80 * precip / maxp) ~ day | month,
         data = SeatacWeather, lty = 1, col = "black",
```

---

[2] As with `type`, this can be supplied directly to `xyplot()`, which will pass it to `panel.superpose()` through `panel.xyplot()`. See `?panel.superpose` for further details.

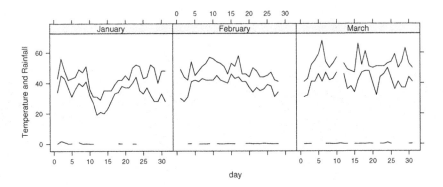

Figure 5.11. Daily meteorological data recorded at the Seattle–Tacoma airport. This figure represents an unsuccessful first attempt to incorporate both rainfall and temperature measurements in a single graphic. The problems arise because the units of rainfall and temperature are different, and the ranges of their numeric values are also different. In addition, **type** = "l" is not quite the right choice for rainfall.

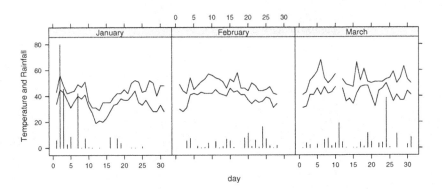

Figure 5.12. Daily rainfall and temperature in Seattle. The rainfall values have been rescaled to make their numeric range comparable to that of the temperature values. The **distribute.type** argument is used to change the interpretation of **type**.

```
ylab = "Temperature and Rainfall",
type = c("l", "l", "h"), distribute.type = TRUE)
```

This still leaves the issue of axis labeling, as Figure 5.12 gives us no information about what the precipitation amounts actually are. A quick-and-dirty solution is to create a fake axis inside using a panel function; the middle panel representing February conveniently has some space on the right that can be used for this purpose. Figure 5.13 is produced by adding a suitable panel function to the previous call.[3]

---

[3] **panel.number()** is a convenient accessor function described in Chapter 12.

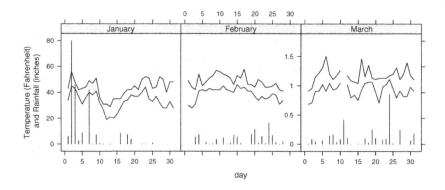

Figure 5.13. A variant of Figure 5.12 that includes a crude axis representing rainfall amounts.

```
> update(trellis.last.object(),
          ylab = "Temperature (Fahrenheit) \n and Rainfall (inches)",
          panel = function(...) {
              panel.xyplot(...)
              if (panel.number() == 2) {
                  at <- pretty(c(0, maxp))
                  panel.axis("right", half = FALSE,
                             at = at * 80 / maxp, labels = at)
              }
          })
```

The techniques outlined in Chapter 8 can be adapted to obtain a more systematic solution, perhaps by having the temperature axis on the left and the rainfall axis on the right. It should be noted, however, that using a common axis to represent multiple units is generally a bad idea, and should be avoided unless there is strong justification.

5.4 Scatter-plot variants for large data

Naïve scatter plots can easily become useless as the number of plotted points increases, causing overplotting. A simple but often effective remedy is to use partially transparent points (as in Figure 3.16); regions with extensive overplotting end up being darker than sparser regions. There are three problems with this solution: not all graphics devices in R support partial transparency, output files in vector formats such as PDF can still end up being large to the point of being impractical, and the solution is not scalable in the sense that with a large enough number of points, overplotting is likely to obscure patterns even with partially transparent points.

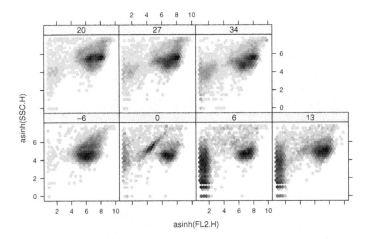

Figure 5.14. A large dataset visualized using hexagonal binning. Each panel visualizes the bivariate distribution of two measurements on cells in blood samples obtained from a blood and marrow transplant patient, taken before and after the transplant. The panels for days 6 and 13 show a large population not seen in the other days.

There are a number of approaches that attempt to deal with this problem, but none are implemented in the default panel function `panel.xyplot()`. In other words, any solution needs to be implemented separately as a custom panel function. One popular approach is to use hexagonal binning (Carr et al., 1987), where the x–y plane is tiled using hexagons which are then colored (or otherwise decorated) to indicate the number of points that fall inside. A panel function implementing this approach is available in the hexbin package (Carr et al., 2006), and can be used to visualize the gvhd10 data encountered in Chapter 3 as follows.

```
> library("hexbin")
> data(gvhd10, package = "latticeExtra")
> xyplot(asinh(SSC.H) ~ asinh(FL2.H) | Days, gvhd10, aspect = 1,
          panel = panel.hexbinplot, .aspect.ratio = 1, trans = sqrt)
```

The result is shown in Figure 5.14. The `asinh()` transformation is largely similar to `log()`, but can handle negative numbers as well. The call is somewhat unwieldy, and can be misleading in the sense that grey levels do not necessarily represent the same number of points in a bin in each panel. A high-level function called `hexbinplot()`, defined in the hexbin package, provides a better interface that addresses this problem and also supports the automatic creation of meaningful legends. An example is given in Figure 14.4.

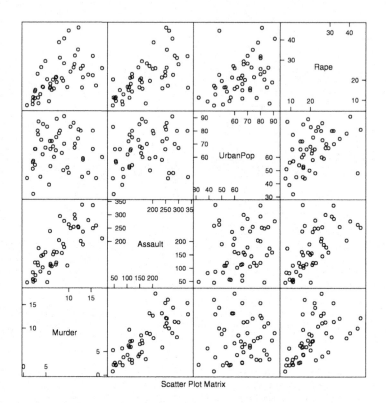

Scatter Plot Matrix

Figure 5.15. A scatter-plot matrix of the `USArrests` data. The `UrbanPop` variable records the percentage of urban population. The remaining variables record the number of arrests per 100,000 population for various violent crimes.

5.5 Scatter-plot matrix

Scatter-plot matrices, produced by `splom()`, are exactly what the name suggests; they are a matrix of pairwise scatter plots given two or more variables. Conditioning is possible, but it is more common to call `splom()` with a data frame as its first argument. Figure 5.15 is a scatter-plot matrix of the `USArrests` dataset, which contains statistics on violent crime rates in the 50 U.S. states in 1973. It is produced by

```
> splom(USArrests)
```

For conditioning with a formula, the primary variables are specified as ~x, where x is a data frame. Figure 5.16 is produced by

```
> splom(~USArrests[c(3, 1, 2, 4)] | state.region,
      pscales = 0, type = c("g", "p", "smooth"))
```

The individual scatter plots are drawn by `panel.splom()`, which is an alias of `panel.xyplot()` and thus honors the same arguments; in particular, it

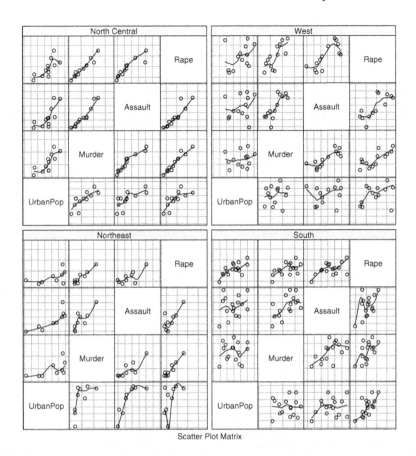

Scatter Plot Matrix

Figure 5.16. Scatter-plot matrices of the `USArrests` data, conditioned on geographical region. The columns have been reordered to make `UrbanPop` the first variable. Reference grids and LOESS smooths have been added as well.

interprets the `type` argument in the same manner. The `pscales` argument is used to suppress the axis labeling. Note that `USArrests` and `state.region` are separate datasets, and can be used together only because they record their data in the same order (alphabetically by state name). The subplots for different levels of `state.region` are slightly separated by default; the amount of separation can be customized using the `between` argument.

The concept of the panel function is somewhat confusing for `splom()`. By analogy with other high-level functions, the panel function should be the one that handles an entire packet (in this case, a conditional data frame subset) and is responsible for the individual scatter plots as well as their layout, including the names of the columns and the axis labeling along the diagonal. In practice, this is instead referred to as the *superpanel function*, and the panel function is the one that renders the individual scatter plots. The superpanel function is specified as the superpanel argument, which defaults to panel.pairs() and is seldom overridden. panel.pairs() allows different panel functions to be used for entries above and below the diagonal, and also allows a user-supplied function for the diagonal blocks. The help page for panel.pairs() describes these and other features in detail. In particular, the pscales and varnames arguments can be used to customize the contents of the diagonal panels relatively easily.

The next example illustrates the use of pscales and varnames. The mtcars dataset (Henderson and Velleman, 1981) records various characteristics of a sample of 32 automobiles (1973–1974 models), extracted from the 1974 *Motor Trend* magazine. Figure 5.17 is a scatter-plot matrix of a subset of the variables recorded, with the number of cylinders as a grouping variable. The varnames argument is used to specify more informative labels for the variables.

```
> splom(~data.frame(mpg, disp, hp, drat, wt, qsec),
        data = mtcars, groups = cyl, pscales = 0,
        varnames = c("Miles\nper\ngallon", "Displacement\n(cu. in.)",
                     "Gross\nhorsepower", "Rear\naxle\nratio",
                     "Weight", "1/4 mile\ntime"),
        auto.key = list(columns = 3, title = "Number of Cylinders"))
```

Note the use of a data argument, where the data frame specified inline in the formula is evaluated. Specifying each variable by name is not always convenient, and one might prefer the equivalent specification

```
> splom(~mtcars[c(1, 3:7)], data = mtcars, groups = cyl)
```

In this case, although groups is evaluated in data, mtcars[c(1, 3:7)] is not. If, as here, there are no conditioning variables, yet another alternative that avoids data altogether is

```
> splom(mtcars[c(1, 3:7)], groups = mtcars$cyl)
```

The appropriate choice in a given situation is largely a matter of taste.

5.5.1 Interacting with scatter-plot matrices

Scatter-plot matrices are useful for continuous multivariate data because they show all the data in a single plot, but they only show pairwise associations and are not particularly helpful in detecting higher-dimensional relationships. However, the layout of the scatter-plot matrix makes it an ideal platform for interactive exploration. In particular, the processes of "linking" and "brushing", where interactively selecting a subset of points in one scatter plot highlights the corresponding points in all the other scatter plots, can be extremely effective in finding hidden relationships. Such interaction with the output produced by splom() is possible, although the capabilities are greatly limited by the underlying graphics system. An example can be found in Chapter 12, which discusses the facilities available for interacting with lattice displays.

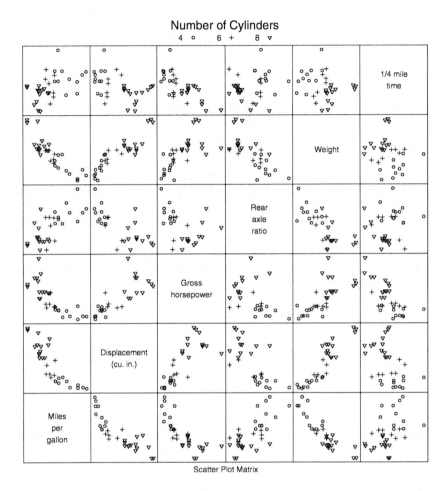

Figure 5.17. A scatter-plot matrix of a subset of the `mtcars` dataset, using the number of cylinders for grouping. As is often the case, using colors (rather than plotting characters) to distinguish between group levels is much more effective. For comparison, a color version of this plot is also available (see color plates).

5.6 Parallel coordinates plot

Like scatter-plot matrices, parallel coordinates plots (Inselberg, 1985; Wegman, 1990) are hypervariate in nature, that is, they show relationships between an arbitrary number of variables. Their design is related to univariate scatter plots; in fact, they are basically univariate scatter plots of all variables of interest stacked parallel to each other (vertically in the implementation in lattice), with values that correspond to the same observation linked by line segments. In other words, the combination of values defining each observation can be decoded by tracing the corresponding "polyline" through the univariate scatter plots for each variable. Parallel coordinates plots can be created using the `parallel()` function in lattice. The primary variable in `parallel()` is a data frame, as in `splom()`, and the formula is interpreted in the same manner. Figure 5.18 shows a parallel coordinates plot of a subset of the columns in the `mtcars` data, using the number of cylinders as a conditioning variable, and the number of carburetors as a grouping variable. The plot is produced by

```
> parallel(~mtcars[c(1, 3, 4, 5, 6, 7)] | factor(cyl),
          mtcars, groups = carb, layout = c(3, 1),
          auto.key = list(space = "top", columns = 3))
```

It is common to scale each variable individually before plotting it, but this can be suppressed using the `common.scale` argument of `panel.parallel()`.

Static parallel coordinates plots, as implemented in lattice, are not particularly useful. They allow pairwise comparisons only between variables that are adjacent. They do not make high-dimensional relationships easy to see; even bivariate relationships between adjacent variables are not always apparent. One point in their favor is that they often make multidimensional clusters easy to see; for example, we can see differences both between panels and between groups in Figure 5.18. This aspect translates to large datasets (if we are careful), as we show in our next example, which is a parallel coordinates plot of the first five columns of one sample in the `gvhd10` dataset. Figure 5.19 is produced by

```
> parallel(~ asinh(gvhd10[c(3, 2, 4, 1, 5)]), data = gvhd10,
          subset = Days == "13", alpha = 0.01, lty = 1)
```

The resulting plot clearly shows multiple high-dimensional clusters; however, the carefully chosen order of variables plays an important role in enabling this "discovery". As with scatter-plot matrices, their hypervariate nature makes parallel coordinates plots ideal candidates for dynamic linking and brushing. Unfortunately, lattice provides no facilities for such manipulation.

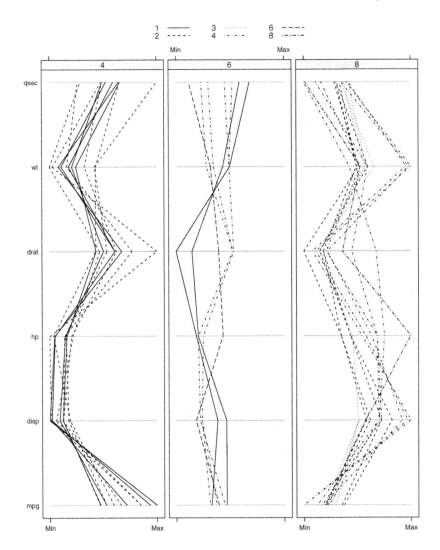

Figure 5.18. A parallel coordinates plot of the `mtcars` data, featuring both conditioning and grouping variables. The groups are not easily distinguishable in this black and white display; color would have been much more effective. Systematic multidimensional differences in the polyline patterns can be seen both between panels and between groups within panels.

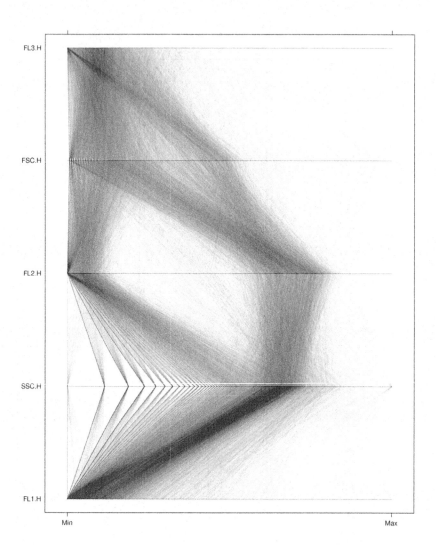

Figure 5.19. Parallel coordinates plot of one sample from the **gvhd10** dataset. The dataset is moderately large, and the display consists of 9540 polylines. The lines are partially transparent, largely alleviating potential problems due to overplotting. This also serves to convey a sense of density, because regions with more line segments overlapping are darker.

6

Trivariate Displays

Trivariate displays encode three primary variables in a panel. There are four high-level functions in lattice that produce trivariate displays: cloud() creates three-dimensional scatter plots of unstructured trivariate data, whereas levelplot(), contourplot(), and wireframe() render surfaces or two-dimensional tables evaluated on a systematic rectangular grid. Of these, cloud() and wireframe() are similar in that they both create two-dimensional projections of three-dimensional constructs, and they share several common arguments that control the details of the projection.

6.1 Three-dimensional scatter plots

We begin with cloud(), which produces three-dimensional scatter plots. Most of the discussion in this section about projection and how to control it in cloud() applies to wireframe() as well. We continue with the quakes example from the previous chapter. In Figure 5.6, we looked at a two-dimensional scatter plot of lat and long, with depth coded by grey level. The natural next step is to look at these in three dimensions. Figure 6.1 is produced by

```
> quakes$Magnitude <- equal.count(quakes$mag, 4)
> cloud(depth ~ lat * long | Magnitude, data = quakes,
        zlim = rev(range(quakes$depth)),
        screen = list(z = 105, x = -70), panel.aspect = 0.75,
        xlab = "Longitude", ylab = "Latitude", zlab = "Depth")
```

As before, we use the shingle Magnitude as a conditioning variable. The first part of the formula has a structure that is different from the ones we have encountered before. It has the form z ~ x * y, where z is the term plotted on the vertical axis, and x and y are plotted on the *x*- and *y*-axes. An equivalent form is z ~ x + y. This interpretation is also shared by the other high-level functions discussed in this chapter.

The cloud() function works by projecting points in three dimensions onto the two-dimensional display area. This is a fairly standard operation, and the

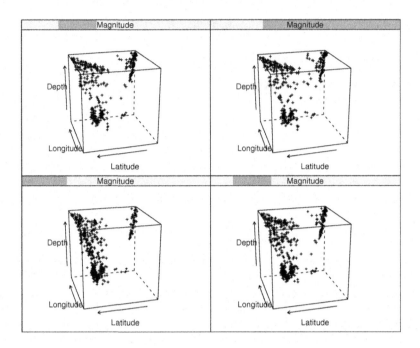

Figure 6.1. A three-dimensional scatter plot of earthquake epicenters in terms of latitude, longitude, and depth. Arrows indicate the direction in which the axes increase; the one for the `depth` is misleading because `zlim` has been reversed. A shingle derived from earthquake magnitude is used as a conditioning variable.

procedure is roughly as follows.[1] The first step is to determine a bounding box in three dimensions. By default, it is defined by the range of the data in each of the dimensions, but this can be changed by the `xlim`, `ylim`, and `zlim` arguments. The data are next centered and scaled, separately for each dimension. The center of the scaled bounding box is the origin, and the lengths of each side are usually the same. The latter can be controlled by the `aspect` argument, which in `cloud()` is a numeric vector of length 2. `aspect[1]` gives the ratio of the length of the scaled bounding box along the y-axis and that along the x-axis. Similarly, `aspect[2]` gives the ratio of lengths along the z- and x-axes. Note that this use of `aspect` is different from the normal use, which is to determine the aspect ratio of the panel. That purpose is served by the `panel.aspect` argument in this case.

The final step is to compute the two-dimensional projection. This is essentially defined by a viewpoint or "camera position" in three-dimensional space, in terms of the scaled coordinate system. Instead of being specified directly,

---

[1] These details are not strictly necessary for casual use, but are helpful in understanding some of the arguments we encounter later.

this viewpoint is determined by two arguments, `screen` and `distance`. `screen` defines the direction of the viewing point with respect to the origin, and `distance` the distance from it, determining the amount of perspective.

The direction is defined as a series of rotations of the bounding box. The viewpoint is initially set to a point on the positive z-axis, so that the positive x-axis points towards the right of the page, the positive y-axis points towards the top, and the positive z-axis is perpendicular to the page pointing towards the viewer. The bounding box, along with the data inside, can be rotated along any of these axes, one at a time, as many times as desired. The rotations are specified through the `screen` argument, which should be a list of named values, with names `x`, `y`, and `z` (each repeated 0 or more times), containing the amount of rotation in degrees. In the example above, we have `screen = list(z = 105, x = -70)`, which means that the bounding box was first rotated 105 degrees along the z-axis, followed by a rotation of -70 degrees along the x-axis. An alternative is to specify a 4×4 transformation matrix `R.mat` in homogeneous coordinates, to be applied to the data before `screen` rotates the view further. We do not go into the details of homogeneous coordinates as they are largely irrelevant; the important thing to know is that it is the de facto standard for specifying three-dimensional transformations and can thus be used to import a transformation from a different projection system. For example, the traditional graphics function `persp()` uses a different set of arguments to define a viewpoint, but its return value is a transformation matrix suitable for use as the `R.mat` argument. Conversely, the `ltransform3dMatrix()` function in lattice computes a suitable transformation matrix given a `screen` specification.

The other component defining the projection is perspective. Projections can be orthogonal, characterized by the feature that lines parallel in three-dimensional space remain parallel in the projection. Such plots can be obtained by setting the `perspective` argument to `FALSE`. Perspective projections are usually preferable, as they are a closer representation of how we view three-dimensional objects; specifically, distant objects are smaller and parallel lines appear to converge at a finite "horizon". The amount of perspective is determined by the `distance` argument, which is inversely related to the distance of the viewpoint from the center of the bounding box. Reasonable values of `distance` are between 0 and 1. Orthogonal projection can be thought of as viewing from an infinite distance,[2] and `distance = 0` is equivalent to `perspective = FALSE`.

We have already seen some of these arguments used in the previous example. We see a couple of new ones in the following call that produces Figure 6.2.

```
> cloud(depth ~ lat * long | Magnitude, data = quakes,
        zlim = rev(range(quakes$depth)), panel.aspect = 0.75,
        screen = list(z = 80, x = -70), zoom = 0.7,
```

---

[2] Through an infinitely powerful telescope that magnifies the view to fit our screen.

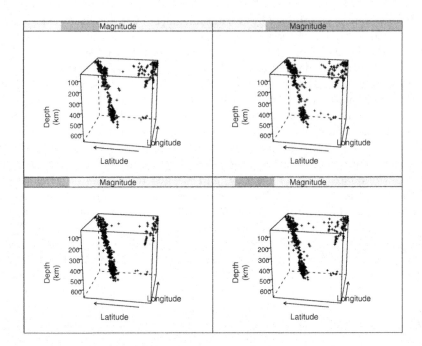

Figure 6.2. Another look at the locations of earthquake epicenters, from a different viewing direction and a few other variations. Together with Figure 6.1, this plot suggests that most of the epicenters are located along one of two distinct planes in three-dimensional space.

```
scales = list(z = list(arrows = FALSE, distance = 2)),
xlab = "Longitude", ylab = "Latitude",
zlab = list("Depth\n(km)", rot = 90))
```

In both examples, `zlim` is specified as an inverted range, so that depth values increase downward rather than upward. The scales are by default annotated by arrows indicating directions of the bounding box axes (which in the case of the z-axis in our example does not match the direction of the data axis). In the second example, we have used the `zoom` argument to shrink the plot slightly to make room for the axis labels, and the `scales` argument to replace the z-axis arrow by labeled tick marks. The default plotting character is a three-dimensional crosshair of sorts, consisting of three intersecting line segments, each parallel to one of the axes. The lengths of the segments are constant in three-dimensional space, but in a perspective projection the projected lengths depend on depth (those closer to the viewer are longer). In theory, this serves as a depth cue, although the benefits are negligible in practice. Other plotting characters can be specified using the `pch` argument, but the perspective transformation is not applied to them.

6.1.1 Dynamic manipulation versus stereo viewing

Projection-based three-dimensional displays benefit greatly from the ability to interactively manipulate details of the projection, such as the viewing direction. Not all features of the data are equally emphasized from all viewpoints, and it is extremely helpful to be able to choose one interactively. Unfortunately, lattice is implemented using a primarily static graphics paradigm, and support for interactive manipulation is sketchy at best. Even non-interactive manipulation, such as producing an animation by systematically moving the viewpoint in small increments, is helpful as the sense of motion it generates is a powerful cue for depth perception. This is possible with lattice in principle, but rendering is currently too slow for it to be practical. When such interaction is desired, alternative visualization systems such as GGobi (Swayne et al., 2003) and the OpenGL-based rgl package can prove to be much more effective unless lattice features such as conditioning are critical.

A couple of simple tricks can alleviate these problems to some extent. To get a comprehensive picture of the data, one can simultaneously view them from several angles. And although motion is not an option for static displays, stereo viewing can be almost as effective, although it does take some getting used to. The basic idea of stereo viewing is to simulate binocular vision by looking at two slightly different pictures through the two eyes; in particular, the one viewed by the right eye should be based on a viewpoint that is slightly to the right of the viewpoint defining the one seen by the left eye. In terms of the interface described above, this means that the "right eye" plot should be rotated clockwise along the y-axis by a small amount.

We combine both these ideas in Figure 6.3. Because the previous two plots suggest no strong dependence of the distribution of epicenters on earthquake magnitudes, we drop the conditioning variable. Our goal is thus to plot a packet containing the same data several times from different viewpoints. One way to implement this is to create separate "trellis" objects for each viewpoint and plot them one by one on the same page. A slightly less obvious approach, used here, is to take advantage of the indexing semantics of "trellis" objects. As we saw in Chapter 2, "trellis" objects can be indexed just as regular R arrays. In particular, an index can be repeated to repeat packets. We start by creating an object containing the data.

```
> p <-
      cloud(depth ~ long + lat, quakes, zlim = c(690, 30),
            pch = ".", cex = 1.5, zoom = 1,
            xlab = NULL, ylab = NULL, zlab = NULL,
            par.settings = list(axis.line = list(col = "transparent")),
            scales = list(draw = FALSE))
```

Next, we repeat it a suitable number of times and update it with a layout and a panel function that chooses a viewpoint depending on the position of the panel in the layout. Figure 6.3 is produced by

```
> npanel <- 4
> rotz <- seq(-30, 30, length = npanel)
> roty <- c(3, 0)
> update(p[rep(1, 2 * npanel)],
         layout = c(2, npanel),
         panel = function(..., screen) {
             crow <- current.row()
             ccol <- current.column()
             panel.cloud(..., screen = list(z = rotz[crow],
                                            x = -60,
                                            y = roty[ccol]))
         })
```

The current row and column are determined inside the panel function using the functions current.row() and current.column(), which we encounter more formally in Chapter 13. Rows in the figure represent different viewing directions, and columns differ by a small (three degrees) rotation along the y-axis. Viewing the result in stereo is somewhat nontrivial, but gets easier after the first time. The trick is to focus the eyes beyond the page, so that the figures on the left and the right column merge together. This process can be catalyzed by using a home-grown stereo viewer; roll up two pieces of paper tightly enough so that only one panel can be seen through each, then use one with each eye to look at different panels.

6.1.2 Variants and panel functions

Just as with other high-level functions, the default panel function in cloud() supports some variants of the standard display shown above, and the option of a user-supplied panel function provides further flexibility. In particular, the groups argument produces grouped displays as usual, and the type argument can be used to join the points by lines (type = "l"). Another useful value of type is type = "h", which causes points to be joined to a "base" plane by vertical lines. Later in this chapter, we show how this feature could be used to create a three-dimensional bar chart of sorts. It can also be useful when absolute (rather than relative) values are being compared, as lengths are easier to compare than position after projection. In the following example, we plot the estimated population density in U.S. states (excluding Alaska and Hawaii) in 1975 as a function of their geographical "center". The data are available in separate R datasets, which we first need to collect together.

```
> state.info <-
      data.frame(name = state.name, area = state.x77[, "Area"],
                 long = state.center$x, lat = state.center$y,
                 population = 1000 * state.x77[, "Population"])
> state.info$density <- with(state.info, population / area)
```

Figure 6.4 is produced by

```
> cloud(density ~ long + lat, state.info,
        subset = !(name %in% c("Alaska", "Hawaii")),
```

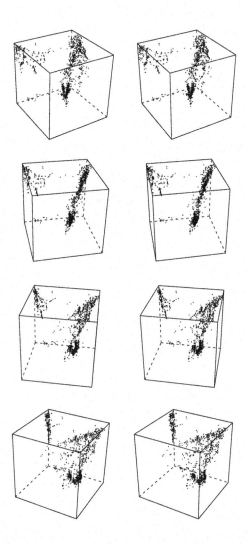

Figure 6.3. Unconditional three-dimensional scatter plots of earthquake epicenters, from different viewing directions. The rows represent different viewpoints, whereas columns differ only by a small rotation along the y-axis, simulating the difference between the positions of the left and right eyes. It is possible to achieve the illusion of depth by focusing the eyes on a point beyond the page and merging the two columns. The effect is often hard to achieve the first time, and it may help to look at the two columns separately through two pieces of rolled-up paper, creating a crude stereo viewer of sorts.

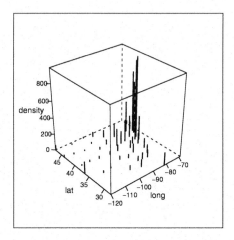

Figure 6.4. Population densities of U.S. states in 1975. Technically, the display is a three-dimensional scatter plot of densities on the z-axis plotted against approximate geographical centers of the states on the x–y plane. Line segments joining the points to their projections on the x–y plane encode the densities by length, making comparison easier. Although the overall spatial pattern is easily identifiable, it is difficult to associate the line segments with individual states. The aspect ratio could also be improved.

```
type = "h", lwd = 2, zlim = c(0, max(state.info$density)),
scales = list(arrows = FALSE))
```

A much more useful version of this plot is given in Figure 6.5, where state boundaries have been added to the bottom plane to serve as a reference. Creating such a plot is not difficult if we have access to state boundary data, but it requires some concepts we have not yet encountered; for this reason, the code to produce Figure 6.5 is postponed until Chapter 13.

6.2 Surfaces and two-way tables

The remaining trivariate functions in lattice are primarily intended for rendering surfaces and other array-like data, where the z-values are evaluated on a regular rectangular grid defined by the x- and y-values. In other words, the z-values form a matrix (at least conceptually), and the x- and y-values represent rows and columns of that matrix. Before going into the details of the individual functions, we discuss situations where they might be appropriate and how to prepare data for use with them.

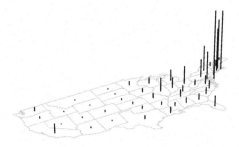

Figure 6.5. An improved version of Figure 6.4. A map of state boundaries on the x–y plane provides a useful visual reference. The aspect ratio is now more natural, and the distracting bounding box has been removed, along with the panel border.

6.2.1 Data preparation

Surfaces are different from other array-like data, as they are in principle smooth, or at least continuous, and can be abstractly represented as a function of two variables. However, they are conveniently represented as matrices containing evaluations of the function on a grid. Tables, on the other hand, are inherently discrete, and two-dimensional tables in particular are naturally represented as matrices. The visualization functions we discuss, namely `wireframe()`, `levelplot()`, and `contourplot()`, do not respect the distinction between surfaces and tables, and the user should be careful to use them in ways suitable for the data.

Before getting to the visualization step, one often has to preprocess the data to get them into a suitable form. We look at some typical examples before using them in plots. The most convenient situation is when the data are already evaluated on a grid, perhaps in the form of a matrix. Our first example, familiar to many R users, is the `volcano` data, which records the elevation of Maunga Whau (Mt. Eden), one of several extinct volcanos in the Auckland region, on a 10 m by 10 m grid. The data are in the form of a matrix, and there are no conditioning variables. Our next example is a correlation matrix, derived from data on car models on sale in the United States in 1993. The data are available in the MASS package.

```
> data(Cars93, package = "MASS")
> cor.Cars93 <-
      cor(Cars93[, !sapply(Cars93, is.factor)], use = "pair")
```

We exclude the categorical variables, although some of them could have been used for conditioning. Our third example is a multiway frequency table, using the Chem97 data again. We create a frequency table of score by gcsescore (discretized into ten equally sized groups) and gender.

```
> data(Chem97, package = "mlmRev")
> Chem97$gcd <-
      with(Chem97,
          cut(gcsescore,
              breaks = quantile(gcsescore, ppoints(11, a = 1))))
> ChemTab <- xtabs(~ score + gcd + gender, Chem97)
```

This of course creates a three-dimensional array, with the third dimension (gender) a natural conditioning variable. As with other lattice functions, it is helpful to convert this to a data frame, to be used with a formula. This can be done using the as.data.frame.table() function.[3]

```
> ChemTabDf <- as.data.frame.table(ChemTab)
```

Our last example is somewhat longer, and involves evaluating fitted regression surfaces on a regular grid. We use the environmental dataset (Bruntz et al., 1974; Cleveland, 1993), which consists of daily measurements of ozone concentration, wind speed, temperature, and solar radiation in New York City for 111 days in 1973. We fit regression models which predict ozone concentration, an indicator of smog, using the other measurements as predictors. As in the original analysis, we use cube root of ozone concentration as the response.

```
> env <- environmental
> env$ozone <- env$ozone^(1/3)
```

For purposes of conditioning, we could also create shingles from the predictors. For example, Radiation is used as a conditioning variable below to create the three-dimensional scatter plots in Figure 6.6.

```
> env$Radiation <- equal.count(env$radiation, 4)
> cloud(ozone ~ wind + temperature | Radiation, env)
```

A scatter-plot matrix is another useful visualization of the data; Figure 6.7 is produced by

```
> splom(env[1:4])
```

We next fit four regression models. The first model is a standard linear regression model. The remaining three are non-parametric; two are variants of LOESS (Cleveland and Devlin, 1988; Cleveland and Grosse, 1991), and the third uses local regression (Loader, 1999) from the locfit package.

```
> fm1.env <- lm(ozone ~ radiation * temperature * wind, env)
> fm2.env <-
      loess(ozone ~ wind * temperature * radiation, env,
```

---

[3] This also works for matrices such as volcano, although they can be used directly as well.

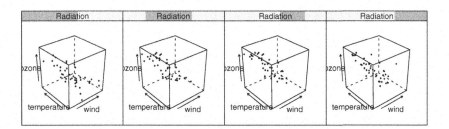

Figure 6.6. Conditional three-dimensional scatter plots showing the relationship among four continuous variables. The fourth variable, radiation, is used for conditioning.

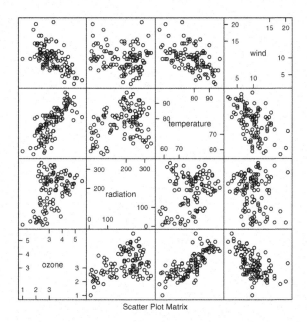

Scatter Plot Matrix

Figure 6.7. A scatter-plot matrix of ozone concentration, radiation, temperature, and wind speed. Neither this plot nor Figure 6.6 fully captures the four-dimensional relationship, but both are useful nonetheless. For our purposes, the most important feature is the correlation in certain pairwise scatter plots. For example, wind speed and temperature are negatively correlated, so there are no observations with high temperature and high wind speed.

```
              span = 0.75, degree = 1)
> fm3.env <-
      loess(ozone ~ wind * temperature * radiation, env,
            parametric = c("radiation", "wind"),
            span = 0.75, degree = 2)
> library("locfit")
> fm4.env <- locfit(ozone ~ wind * temperature * radiation, env)
```

Our eventual goal is to display the fitted regression surfaces. To do so, we first need to evaluate the predicted ozone concentrations on a regular grid of predictor values. There are three predictors, and we can only use two to define a surface, we therefore use one as a conditioning variable. We first create the vectors of values for each predictor that define the margins of the grid.

```
> w.mesh <- with(env, do.breaks(range(wind), 50))
> t.mesh <- with(env, do.breaks(range(temperature), 50))
> r.mesh <- with(env, do.breaks(range(radiation), 3))
```

The `expand.grid()` function can construct a full grid, in the form of a data frame, from these margins.

```
> grid <-
      expand.grid(wind = w.mesh,
                  temperature = t.mesh,
                  radiation = r.mesh)
```

The final step is to add columns in this data frame representing each of the fitted models. This can be easily done using the `predict()` methods for each of the fits.

```
> grid[["fit.linear"]] <- predict(fm1.env, newdata = grid)
> grid[["fit.loess.1"]] <- as.vector(predict(fm2.env, newdata = grid))
> grid[["fit.loess.2"]] <- as.vector(predict(fm3.env, newdata = grid))
> grid[["fit.locfit"]] <- predict(fm4.env, newdata = grid)
```

We now use these examples to create some plots.

6.2.2 Visualizing surfaces

We begin with the last example. Figure 6.8 is created with

```
> wireframe(fit.linear + fit.loess.1 + fit.loess.2 + fit.locfit ~
                                    wind * temperature | radiation,
            grid, outer = TRUE, shade = TRUE, zlab = "")
```

As with `cloud()`, the formula has the form z ~ x * y, but it is assumed in this case that x and y define a regular grid. Wind speed and temperature are used here as the x and y variables, and radiation as a conditioning variable. In this example, the formula actually contains four z variables separated by + signs. Normally, these would be used for grouping within each panel (as explained in Chapter 10), but the outer = TRUE argument causes them to be used for conditioning. By plotting all the fits together, we can compare

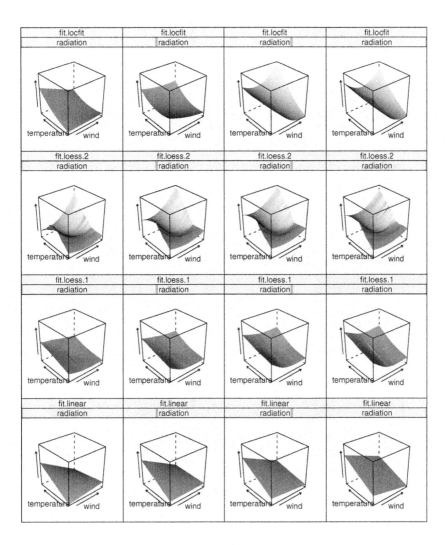

Figure 6.8. Wireframe plots of the fitted regression surfaces. Rows represent four different regression models and columns represent four levels of radiation; each panel graphs the surface representing predicted ozone concentration as a function of temperature and wind speed for a fixed level of radiation. The four models give widely inconsistent results when wind and temperature are both low or both high (the corners closest to and farthest from the viewer); these are regions where there are few actual observations.

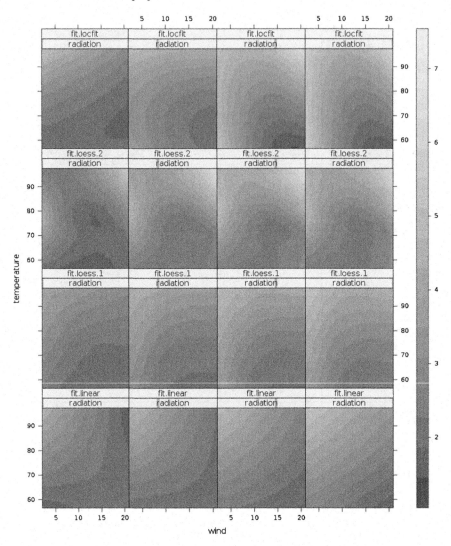

Figure 6.9. False-color level plots of fitted regression surfaces, in the same layout as Figure 6.8. This representation is independent of viewing direction, and easily conveys relative order between two points. Magnitudes of changes are harder to interpret without constantly referring to the color key. The choice of color can be important; in particular, greyscale gradients can only change in one direction, whereas true color gives more options. See the color plates for a color version of this figure.

their global characteristics. In each case, the predicted ozone levels generally increase with radiation. What is different is the behavior of the fitted surfaces as both wind speed and temperature increase. The reason for this can be understood if we look back at Figures 6.6 and 6.7; there are practically no observed data points with high values of both wind speed and temperature, and thus any regression fit will be unreliable in this region.

The `levelplot()` function has an interface identical to that of `wireframe()`, and works on the same type of data. However, instead of using projections, it uses a false-color gradient to encode the z variable. Figure 6.9 presents the same data as Figure 6.8 and is created by

```
> levelplot(fit.linear + fit.loess.1 + fit.loess.2 + fit.locfit ~
                          wind * temperature | radiation,
          data = grid)
```

Yet another function with the same interface is `contourplot()`, which instead of using colors, draws contour lines. Generally, the contours are labeled by the level (value of the z variable) they represent, which may be preferable if the exact values are important; the disadvantage to this approach is that one cannot use too many levels, as then the labels tend to overlap. Figure 6.10 is created by

```
> contourplot(fit.locfit ~ wind * temperature | radiation,
             data = grid, aspect = 0.7, layout = c(1, 4),
             cuts = 15, label.style = "align")
```

All three functions have methods that work directly on a matrix. The following calls illustrate their use with the `volcano` data. The resulting plots are shown together in Figure 6.11.

```
> levelplot(volcano)
> contourplot(volcano, cuts = 20, label = FALSE)
> wireframe(volcano, panel.aspect = 0.7, zoom = 1, lwd = 0.5)
```

Note that the default of the `aspect` argument is different for these methods.

6.2.3 Visualizing discrete array data

Our other examples, a correlation matrix and a frequency table, represent data that are discrete in nature. `wireframe()` and `contourplot()`, which are designed for continuous surfaces, should not be used for such data. However, `levelplot()` can still be used, as we do in Figure 6.12, where grey levels are used to represent pairwise correlations between various continuous characteristics of several passenger car models for sale in the United States in 1993. Plots of correlation matrices are similar to scatter-plot matrices in structure, with individual scatter plots replaced by scalar summaries, namely the correlations. Figure 6.12 is produced by

```
> levelplot(cor.Cars93,
          scales = list(x = list(rot = 90)))
```

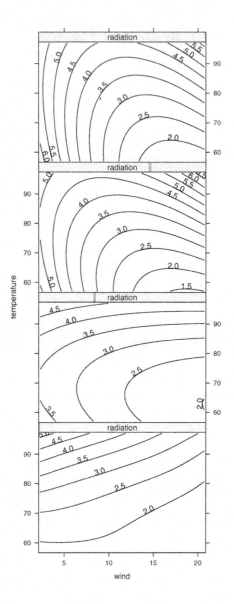

Figure 6.10. Contour plots of fitted regression surfaces. The design is similar to level plots, but shows the boundaries between levels rather than the levels themselves (although both can be combined in a single display). Contours can be labeled with the values they represent, enabling more direct decoding of the z variable. The density (closeness) of contour lines gives a sense of how fast the surface changes.

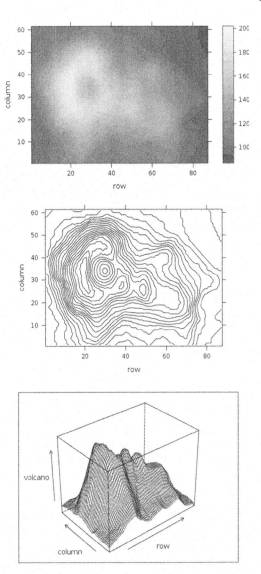

Figure 6.11. Visualizations of the topography of Mt. Eden, an extinct volcano in the Auckland region. The elevation values are stored as a matrix in the `volcano` dataset, which is used in the *"matrix"* methods for `levelplot()`, `contourplot()`, and `wireframe()` to produce the displays here.

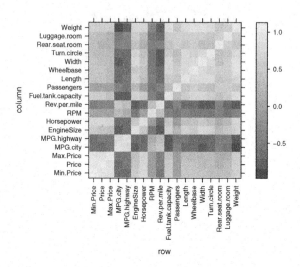

Figure 6.12. A correlation matrix derived from the `Cars93` data, visualized as a false-color image with grey levels encoding correlation. It would have been useful to be able to detect zero correlation easily, but this is not possible using grey levels alone. In addition, the order of rows and columns is arbitrary, making it difficult to see any patterns.

In this example, the order of rows and columns is arbitrary, and as with other types of plots, reordering them in a systematic manner can be helpful. One simple way to reorder rows or columns of a matrix is to first cluster them, and then order them in a manner consistent with the clustering. In the following example, we do so using the `hclust()` function. We also specify an explicit vector of levels where the colors change, instead of using the range of the correlations.

```
> ord <- order.dendrogram(as.dendrogram(hclust(dist(cor.Cars93))))
> levelplot(cor.Cars93[ord, ord], at = do.breaks(c(-1.01, 1.01), 20),
            scales = list(x = list(rot = 90)))
```

The resulting plot is shown in Figure 6.13. Other displays of correlation matrices, such as those described by Friendly (2002), can be produced from similar data using custom panel functions; two examples can be seen in Figures 13.5 and 13.6.

The frequency table derived from the `Chem97` data can be similarly visualized using `levelplot()`. It is often helpful to encode frequencies after taking their square root. To do so with a color gradient, we must also modify the color key to reflect this transformation. Figure 6.14 is produced by

```
> tick.at <- pretty(range(sqrt(ChemTabDf$Freq)))
> levelplot(sqrt(Freq) ~ score * gcd | gender, ChemTabDf,
```

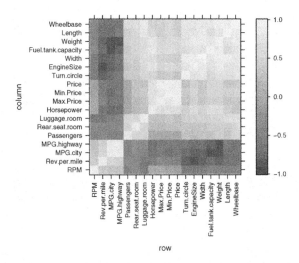

Figure 6.13. A slightly modified version of Figure 6.12, with rows and columns reordered according to a hierarchical clustering. Similar columns are now easily identifiable. This does not address the problem of emphasizing strength and direction of correlation separately, which is considered later in Chapter 13.

```
shrink = c(0.7, 1), aspect = "iso", colorkey =
    list(labels = list(at = tick.at, labels = tick.at^2)))
```

In addition to a color gradient, this example also encodes the z-values using the size of the rectangles. The details are controlled by `shrink`, which is an argument of the default panel function `panel.levelplot()`.

Although `wireframe()` is unsuitable for discrete data, they can still be plotted in a three-dimensional projection using `cloud()`, ignoring the regular structure in the x- and y-values. For example, the following code would create a three-dimensional bar chart of sorts where frequencies are encoded by line segments.

```
> cloud(Freq ~ score * gcd | gender, data = ChemTabDf, type = "h",
        aspect = c(1.5, 0.75), panel.aspect = 0.75)
```

We do not show the results of this call, but instead use the `panel.3dbars()` function available in the latticeExtra package to create more "solid" versions of the bars. Figure 6.4 is produced by

```
> library("latticeExtra")
> cloud(Freq ~ score * gcd | gender, data = ChemTabDf,
        screen = list(z = -40, x = -25), zoom = 1.1,
        col.facet = "grey", xbase = 0.6, ybase = 0.6,
        par.settings = list(box.3d = list(col = "transparent")),
        aspect = c(1.5, 0.75), panel.aspect = 0.75,
        panel.3d.cloud = panel.3dbars)
```

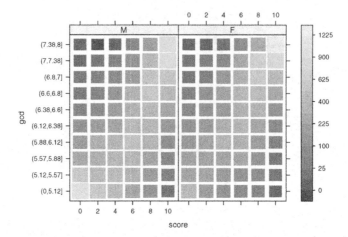

Figure 6.14. False color plots of a table derived from the `Chem97` data, showing the relationship between `score` and `gcsescore`, conditioning on `gender`. Counts are encoded by grey level as well as size of the rectangles. As with Figure 6.9, using color instead of grey levels considerably increases the usefulness of the display; in particular, the rectangles on the lower-left and upper-right corners are almost the same color as the background, making them difficult to see.

Note that we have specified a `panel.3d.cloud` argument rather than a `panel` argument; this is because the panel function in `cloud()` and `wireframe()` are responsible for computing the projections and drawing the bounding box, a task we normally wish to leave unchanged. The data-dependent part of the display is the responsibility of the `panel.3d.cloud` argument (`panel.3d.wireframe` for `wireframe()`) of `panel.cloud()`, and this is the piece we replace in our example.

6.3 Theoretical surfaces

The methods we used to plot regression surfaces using `wireframe()` can be easily adapted to mathematical surfaces. For our next example, we consider four bivariate copulas (Nelsen, 1999), which are essentially joint distributions on the unit square with uniform marginals. Our goal is to plot the corresponding density functions, as computed by the `dcopula()` function in the copula package (Yan and Kojadinovic, 2007). We start, as before, by defining a grid and adding columns to it:

```
> library("copula")
> grid <-
      expand.grid(u = do.breaks(c(0.01, 0.99), 15),
                  v = do.breaks(c(0.01, 0.99), 15))
```

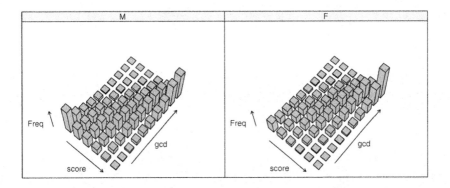

Figure 6.15. A three-dimensional bar chart showing the same data as Figure 6.14. The usefulness of such plots, compared to level plots, is questionable, as the information perceived depends to a considerable extent on choices that are not data-related, such as the viewing direction.

```
> grid$frank  <- with(grid, dcopula(frankCopula(2),   cbind(u, v)))
> grid$gumbel <- with(grid, dcopula(gumbelCopula(1.2), cbind(u, v)))
> grid$normal <- with(grid, dcopula(normalCopula(.4),  cbind(u, v)))
> grid$t      <- with(grid, dcopula(tCopula(0.4),      cbind(u, v)))
```

Figure 6.16 is now produced by

```
> wireframe(frank + gumbel + normal + t ~ u * v, grid, outer = TRUE,
            zlab = "", screen = list(z = -30, x = -50), lwd = 0.5)
```

The densities appear almost flat, as the vertical axis is dominated by changes close to the corners, even though we left out the corners themselves. In Figure 6.17, we try plotting the log-transformed densities, with better results.

```
> wireframe(frank + gumbel + normal + t ~ u * v, grid, outer = TRUE,
            zlab = "", screen = list(z = -30, x = -50),
            scales = list(z = list(log = TRUE)), lwd = 0.5)
```

Instead of transforming each term in the formula separately, we use the `scales` argument, which is described in detail in Chapter 8. Needless to say, these surfaces can also be visualized using `levelplot()`.

6.3.1 Parameterized surfaces

The surfaces we have seen thus far are defined as $z = f(x, y)$, where x and y vary over a continuous interval, approximated by a discrete grid. A more

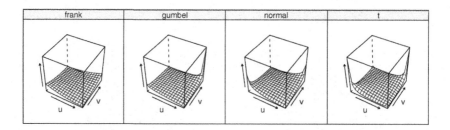

Figure 6.16. A wireframe plot representing the probability density function of four copulas. The surfaces appear to be largely flat because some of the densities increase rapidly close to the corners.

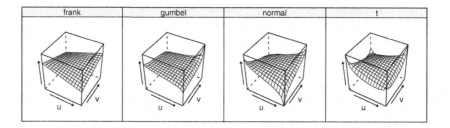

Figure 6.17. Figure 6.16 rerendered with log densities on the z-axis. The surfaces are now much easier to compare.

general way of representing surfaces is to parameterize them as functions of the form

$$f : [\,0, 1\,] \times [\,0, 1\,] \longrightarrow \mathbb{R}^3$$

where every point on the surface corresponds to a point (u, v) on the unit square, with coordinates in three-dimensional space given by

$$f(u, v) = (x(u, v), y(u, v), z(u, v))$$

A simple example of a parameterized surface is a sphere, which can be represented by the equations

$$x(\theta, \phi) = \cos \theta \cos \phi$$
$$y(\theta, \phi) = \sin \theta \cos \phi$$
$$z(\theta, \phi) = \sin \phi$$

where $\theta, \phi \in [-\pi, \pi]$ can be thought of as longitude and latitude, respectively (we use this interpretation later to create Figure 13.9). The domain here is not the unit square, but this can be easily rectified by a simple scale and location shift. A somewhat more complicated, but fairly well-known example is the "figure 8" immersion of the Klein bottle, with a possible parameterization given by

$$x = \cos u \ \left(r + \cos\frac{u}{2} \cdot \sin tv - \sin\frac{u}{2} \cdot \sin 2tv\right)$$

$$y = \sin u \ \left(r + \cos\frac{u}{2} \cdot \sin tv - \sin\frac{u}{2} \cdot \sin 2tv\right)$$

$$z = \sin\frac{u}{2} \cdot \sin tv + \cos\frac{u}{2} \cdot \sin tv$$

with $u, v \in [\,0, 2\pi]$, where r controls the thickness of the loops, and t gives the number of twists.

One interesting (although of little value in practical data analysis) feature of `wireframe()` is that it can draw parameterized surfaces. Such plots are created using the familiar formula z ~ x * y, but require x, y, and z to be all matrices with the same dimensions, representing coordinates of the parameterized surface evaluated over a grid of (u, v)-values. The following sequence of calls sets up the pieces required to create such matrices for the parameterization given above.

```
> kx <- function(u, v)
      cos(u) * (r + cos(u/2) * sin(t*v) - sin(u/2) * sin(2*t*v))
> ky <- function(u, v)
      sin(u) * (r + cos(u/2) * sin(t*v) - sin(u/2) * sin(2*t*v))
> kz <- function(u, v)
      sin(u/2) * sin(t*v) + cos(u/2) * sin(t*v)
> n <- 50
> u <- seq(0.3, 1.25, length = n) * 2 * pi
> v <- seq(0, 1, length = n) * 2 * pi
> um <- matrix(u, length(u), length(u))
> vm <- matrix(v, length(v), length(v), byrow = TRUE)
> r <- 2
> t <- 1
```

Figure 6.18 is now created with

```
> wireframe(kz(um, vm) ~ kx(um, vm) + ky(um, vm), shade = TRUE,
            screen = list(z = 170, x = -60),
            alpha = 0.75, panel.aspect = 0.6, aspect = c(1, 0.4))
```

6.4 Choosing a palette for false-color plots

Level plots encode a quantitative variable by using a color gradient (or grey levels) to represent numeric values. It is common to include a color key that maps the colors to the values they represent, but one should not expect to be able to use it to make accurate quantitative judgments. Rather, the primary usefulness of level plots is in judging patterns in the variability. A good choice of colors is often critical in how well a particular display serves this purpose.

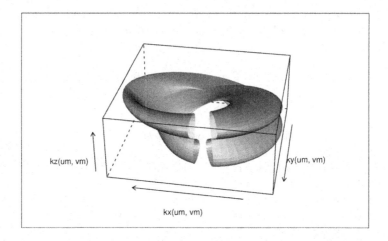

Figure 6.18. A shaded wireframe plot of the "figure 8" immersion of the Klein bottle, created using the parameterized form given in the text. The name comes from the interpretation of the object as a Möbius strip with the usual cross-section (a line segment) replaced by a double loop (like the number 8).

As a case in point, consider the `USAge.df` dataset available in the latticeExtra package, which records estimated population by age and sex in the United States between 1900 and 1979.

```
> data(USAge.df, package = "latticeExtra")
> str(USAge.df)
'data.frame':   12000 obs. of  4 variables:
 $ Age       : num  0 1 2 3 4 5 6 7 ...
 $ Sex       : Factor w/ 2 levels "Male","Female": 1 1 1 1 1 1 1 1 ...
 $ Year      : num  1900 1900 1900 1900 1900 1900 1900 1900 ...
 $ Population: num  0.919 0.928 0.932 0.932 0.928 0.921 0.911 0.899 ..
```

The dataset is large, with interesting local features. We look at some subsets of the data later in Chapter 10; here we consider a visualization of the full dataset using a level plot. In the following call, we use a gradient that is derived from a color palette designed by Cynthia Brewer (Harrower and Brewer, 2003; Neuwirth, 2007).

```
> library("RColorBrewer")
> brewer.div <-
      colorRampPalette(brewer.pal(11, "Spectral"),
                       interpolate = "spline")
> levelplot(Population ~ Year * Age | Sex, data = USAge.df,
            cuts = 199, col.regions = brewer.div(200),
            aspect = "iso")
```

The result, shown in Figure 6.19 along with other color plates, contains a small fluctuation around 1918 for males aged 22 or thereabouts. Unfortunately, neither the default black and white theme nor the default color theme will work well to highlight this feature. The palette used is by no means the only suitable choice; for example, the gradient produced by `terrain.colors()` also performs well in this case. The important point here is not that certain schemes are better than others, rather, it is that different color gradients emphasize different ranges of the data. One should keep this fact in mind when using color to encode numeric values.

Part II

Finer Control

7

Graphical Parameters and Other Settings

In the second part of this book, we take a detailed look at features that are common to all high-level lattice functions, providing a uniform interface to control their output. We start, in this chapter, by describing the system of user settable graphical parameters and other global options.

Graphical parameters are often critical in determining the effectiveness of a plot. Such parameters include obvious ones such as colors, symbols, line types, and fonts for the various elements of a graph, as well as more subtle ones such as the length of tick marks or the amount of space separating different components of the graph. The parameters used in lattice displays are highly customizable. Many of them can be controlled directly by specifying suitable arguments in a high-level function call. Most derive their default values from a system of common global settings that can also be modified by the user. The latter approach has two primary benefits: it allows good global defaults to be specified, and it provides a consistent "look and feel" to lattice graphics while letting the user retain ultimate control.

Not all parameters of interest are graphical. For example, a user may dislike the default argument value `as.table = FALSE` (which orders panels starting from the lower-left corner rather than the upper-left one), and wish to change the default globally rather than specify an additional argument in every call. Several such non-graphical parameters can be customized, through a slightly different system of global options. Both these systems are discussed in this chapter.

7.1 The parameter system

In this section, we present some essential background information about the graphical parameter system. The parameters that are actually available for use and their effect are detailed in the next section.

7.1.1 Themes

Choosing good graphical parameters is a nontrivial task. For grouped displays in particular, one needs colors, plotting characters, line types, and so on, that mesh well together, yet are distinctive enough that each stands out from the others. Furthermore, a choice of colors that is good for points and lines may not be good as a fill color (e.g., for bar charts). The settings system in lattice allows the user to specify a coherent collection of graphical parameters, presumably put together by an expert,[1] to be used as a common source consistently across all high-level plots. Such collections of parameters are henceforth referred to as *themes*.

Unfortunately, it is even harder to find a single theme that is optimal for all display media, to say nothing of individual tastes. Color is vastly more effective than plotting characters or line types in distinguishing between groups; however, black and white printing is often considerably cheaper. Even when available, a good choice of colors for printing may not be as good for viewing on a computer monitor or an overhead projector as these involve fundamentally different physical mechanisms; colors in print are produced by subtracting basic colors, whereas color on monitors and projectors is produced by adding basic colors. Furthermore, the same specification may produce different actual colors on different hardware, sometimes because of the hardware settings, sometimes simply because of differences in the hardware.

7.1.2 Devices

In traditional S graphics, no special consideration was given to the target media. Following the original Trellis implementation in S, lattice attempts to rectify this situation, although not with unqualified success, by allowing graphical settings to be associated with specific devices. As the reader should already know, R can produce graphics on a number of output devices. Each supported platform has a native screen device,[2] as well as several file-based devices. The latter include vector formats such as PDF, Postscript®, SVG (scalable vector graphics), and EMF (on Windows), as well as bitmap formats such as JPEG and PNG. lattice allows a different theme to be associated with each of these devices.

Unfortunately, this does not really solve the problem of settings specific to target media. It is common for PDF documents to be viewed on a screen or projected (especially presentation slides) as well as printed. It is also fairly common practice to print a graphic displayed on the screen using dev.print() and related functions, often via a GUI menu, which simply recreates the graph

---

[1] We do not discuss how one might design an effective collection of settings, as that is beyond the scope of this book. See Ihaka (2003) for a helpful discussion of colors in presentation graphics. The packages RColorBrewer and colorspace provide some useful tools for working with color.

[2] Typically one of x11, quartz, and windows.

without changing the settings to match the target device. Thus, the availability of device-specific settings is only beneficial if the user is disciplined enough, and such settings are possibly confusing for the casual user. For this reason, all devices currently use a common color theme by default, with the exception of postscript, which uses a black and white theme. It is possible for the user to associate other themes as the default for specific devices, and a procedure to do so is outlined later in this chapter. Even if one is not interested in device-specific themes, it is helpful to keep the preceding discussion in mind when reading the rest of this section.

7.1.3 Initializing a graphics device

In traditional R graphics, devices are initialized by calling the corresponding device function (e.g., pdf(), png(), etc.). If a plotting function is called with no open device, the one specified by getOption("device") (typically the native screen device) is opened automatically. This works for lattice plots as well, as long as one is content using the default theme. However, for finer control over the theme used, it is more convenient to initialize a device through the wrapper function trellis.device(). It has the following arguments.

device

This argument determines the device that will be opened (unless new = FALSE). It can be specified either as a device function (e.g., pdf) or as the name of such a function (e.g., "pdf"). By default, getOption("device") is used.

color

Every device has a default theme associated with it, which can be modified fully or partially via the theme argument described below, or after the device is opened. This default theme can be one of two choices, one color and one black and white. The color argument is a logical flag that determines this choice; it defaults to FALSE for postscript devices, and to TRUE for all others.

theme

This argument allows modifications to the default theme to be specified. Details are given below. This argument defaults to lattice.getOption("default.theme").

new

This is a logical flag indicating whether a new device should be initialized. It only makes sense to set this to FALSE when one wishes to reset the currently active device's theme to the settings determined by other arguments. An alternative is to use the trellis.par.set() function described later.

`retain`

> This is also a logical flag. Once a device is open, its settings can be modified. When another instance of the same device is opened later using `trellis.device()`, the settings for that device are usually reset to its defaults. This can be prevented by specifying `retain = TRUE`. Note that settings for different devices are always treated separately, that is, opening a `postscript()` device does not alter the `pdf()` settings (but it does reset the settings of any `postscript` device that is already open).

A theme (as in the `theme` argument above) can be specified either as a list containing parameter values, or a function that produces such a list when called. The structure of the list is discussed a little later. If `theme` is a function, it will not be supplied any arguments, but the device is guaranteed to be open when it is called, so one may use the `.Device` variable inside the function to ascertain what device has been opened. Note that `theme` only modifies the theme determined by other arguments, and need not contain all possible parameters.

One important difference between calling a device function such as `pdf()` directly, and calling it through `trellis.device()`, is that the latter resets the device theme to the initial defaults, undoing any prior changes (unless `retain = TRUE`). This is often the easiest way to recover from experiments with settings that have gotten out of hand.

7.1.4 Reading and modifying a theme

Most elements of a lattice graphic can be controlled by some theme parameter. Of course, one must know which one for this fact to be useful. Once a device is open, the theme associated with it can be queried and modified using the functions `trellis.par.get()` and `trellis.par.set()`. This is best illustrated by an example. Consider Figure 7.1, which gives an alternative visualization of the `VADeaths` dataset, similar to Figure 4.2. The plot is produced by

```
> vad.plot <-
      dotplot(reorder(Var2, Freq) ~ Freq | Var1,
              data = as.data.frame.table(VADeaths),
              origin = 0, type = c("p", "h"),
              main = "Death Rates in Virginia - 1940",
              xlab = "Number of deaths per 100")
> vad.plot
```

Because the absolute rates are encoded by a line "dropping" down to the origin, the light grey reference lines are now somewhat redundant. Let us try, as an exercise, to remove them from the graph. The parameters in a theme are generally identified by names descriptive of their use, and the parameters of the reference line happen to be determined by the settings named `"dot.line"`.

```
> dot.line.settings <- trellis.par.get("dot.line")
> str(dot.line.settings)
```

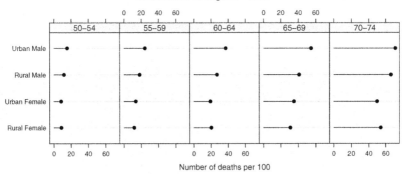

Figure 7.1. A dot plot of death rates in Virginia in the year 1940 across population groups, conditioned on age groups. The rates are encoded by length as well as position, through line segments joining the points to the origin.

```
List of 4
 $ alpha: num 1
 $ col  : chr "#E6E6E6"
 $ lty  : num 1
 $ lwd  : num 1
```

As the output of `str()` suggests, the result is a list of graphical parameters[3] that control the appearance of the reference lines. The simplest way to omit the reference lines is to make them transparent. This can be done by modifying the `dot.line.settings` variable and then using it to change the settings:

```
> dot.line.settings$col <- "transparent"
> trellis.par.set("dot.line", dot.line.settings)
```

While we are at it, let us also double the thickness of the lines being shown, whose parameters are obtained from the `"plot.line"` settings:

```
> plot.line.settings <- trellis.par.get("plot.line")
> str(plot.line.settings)
List of 4
 $ alpha: num 1
 $ col  : chr "#000000"
 $ lty  : num 1
 $ lwd  : num 1
```

---

[3] Where possible, lattice follows the standard naming conventions for graphical parameters: `col` for color, `lty` for line type, `lwd` for line width, `pch` for plotting character, and `cex` for character size. Fonts can be controlled by `font`, or `fontface` and `fontfamily` for finer control. In addition, `alpha` is used for partial transparency (often referred to as alpha-channel transparency for historical reasons) on devices that support it. See the `?par` help page for further details, including valid ways to specify color and line type.

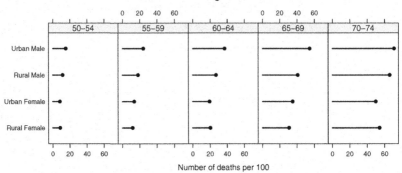

Figure 7.2. Death rates in Virginia. An alternative version of Figure 7.1 with a slightly modified theme. The reference lines, which are mostly redundant, have been removed, and the widths of the line segments have been doubled.

```
> plot.line.settings$lwd <- 2
> trellis.par.set("plot.line", plot.line.settings)
```

We can now simply replot the previously saved object to produce Figure 7.2.

```
> vad.plot
```

An alternative solution that does not require the settings to be modified is to write a suitable panel function. Even though this is not necessary in this example, it is instructive as an illustration of how theme parameters might provide defaults in a panel function.

```
> panel.dotline <-
      function(x, y,
               col = dot.symbol$col, pch = dot.symbol$pch,
               cex = dot.symbol$cex, alpha = dot.symbol$alpha,
               col.line = plot.line$col, lty = plot.line$lty,
               lwd = plot.line$lwd, alpha.line = plot.line$alpha,
               ...)
  {
      dot.symbol <- trellis.par.get("dot.symbol")
      plot.line <- trellis.par.get("plot.line")
      panel.segments(0, y, x, y, col = col.line, lty = lty,
                     lwd = lwd, alpha = alpha.line)
      panel.points(x, y, col = col, pch = pch,
                   cex = cex, alpha = alpha)
  }
```

This panel function explicitly draws the line segments and points to create the display. Thanks to lazy evaluation, the default parameters are obtained from the theme active when the panel function is called. This panel function can now be used in a call such as

```
> update(vad.plot, panel = panel.dotline)
```

It is left as an exercise to the reader to verify that the thickness of the lines in the resulting plot depends on whether the "plot.line" settings were modified beforehand.

7.1.5 Usage and alternative forms

Both trellis.par.get() and trellis.par.set() apply to the theme associated with the currently active device. trellis.par.get(), called with a name argument, returns the associated parameters as a list. When called without a name argument, it returns the full list of settings. trellis.par.set() can be called analogously with arguments name and value, as shown above. However, this is not its only valid form. More than one parameter can be set at once as named arguments, so the two trellis.par.set() calls earlier can be replaced by the single call

```
> trellis.par.set(dot.line = dot.line.settings,
                   plot.line = plot.line.settings)
```

In fact, the replacements may be "incomplete", in the sense that only components being modified need to be supplied. In other words, the above is equivalent to

```
> trellis.par.set(dot.line = list(col = "transparent"),
                   plot.line = list(lwd = 2))
```

Finally, any number of parameters can be supplied together as a list, for example,

```
> trellis.par.set(list(dot.line = list(col = "transparent"),
                       plot.line = list(lwd = 2)))
```

This last option is a convenient way to specify a complete user-defined theme, that is, a subcollection of parameters that provides an alternative look and feel. This is in fact the form that the theme argument in trellis.device() must take when it is a list, and the same applies to its return value when theme is a function.

7.1.6 The par.settings argument

As noted above, trellis.par.set() modifies the current theme. Often, one wants to associate specific parameter values with a particular call rather than globally modify the settings. This can be achieved using the par.settings argument in any high-level lattice call. Whenever the resulting object is plotted, whether immediately or later with a different theme active, these settings are temporarily in effect for the duration of the plot, after which the settings revert to whatever they were before. For example, the following will re-create Figure 7.2 with or without the earlier calls to trellis.par.set().

```
> update(vad.plot,
        par.settings = list(dot.line = list(col = "transparent"),
                            plot.line = list(lwd = 2)))
```

This paradigm is particularly useful, in conjunction with the `auto.key` argument, for grouped displays with non-default graphical parameters. The convenience function `simpleTheme()` can often be used to create a suitable value for `par.settings` with little effort.

7.2 Available graphical parameters

As explained in the previous section, the graphical parameter system can be viewed as a collection of named settings, each controlling certain elements in lattice displays. To take advantage of the system, either by modifying themes or by using them as defaults in custom panel functions, the user must know the names and structures of the settings available. The full list is subject to change, but the most current list can always be obtained by inspecting the contents of a theme, for example, using

```
> names(trellis.par.get())
```

Most of these settings have a common pattern: their value is simply a list of standard graphical parameters such as `col`, `pch`, and so on. Figure 7.3 lists these settings along with their component parameters.

These settings can be broadly divided into two types based on their purpose. Some are intended to control elements common to most lattice displays. These include

par.xlab.text, par.ylab.text, par.main.text, par.sub.text
: which control the various labels (in addition, `par.zlab.text` controls the *z*-axis label in `cloud()` and `wireframe()`),

strip.background, strip.shingle, strip.border
: which control certain aspects of the strips through the default strip function `strip.default()`, and

axis.text, axis.line
: which control the appearance of axes.

Other settings are meant for use by panel functions. Some of these have very specific targets; for example,

box.dot, box.rectangle, box.umbrella
: are used by `panel.bwplot()`,

dot.line, dot.symbol
: are used by `panel.dotplot()`,

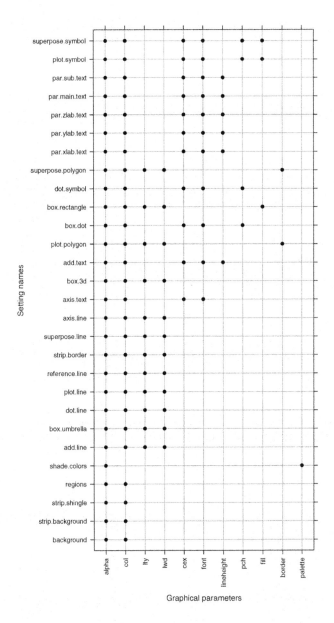

Figure 7.3. The standard graphical settings (at the time of writing). Each setting has a specific purpose, and consists of one or more graphical parameters. The parameter names (`col`, `pch`, etc.) follow the usual R conventions for the most part. The **fontface** and **fontfamily** parameters may be used for finer control over fonts wherever **font** is allowed (see **?gpar** in the grid package).

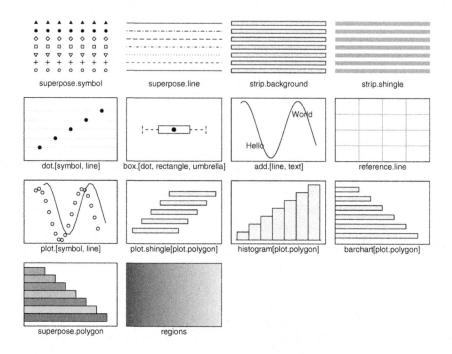

Figure 7.4. A graphical summary of the black and white theme used throughout this book, as produced by **show.settings()**. The other primary built-in parameter scheme available in lattice is the color scheme used for the color plates, which is also the default on all screen devices. Most use of color can be justified by one of two purposes; first, to distinguish data driven elements from non-data elements such as axes, labels and reference lines, and second, to distinguish between levels of a grouping variable in superposed displays. In the default black and white theme, the first goal is largely ignored, and the second is achieved using different symbols and line types (and grey levels when necessary). A summary of the default color theme is shown in the color plates.

plot.line, plot.symbol
  are used (for the most part) by **panel.xyplot()**, **panel.densityplot()**, and **panel.cloud()**,

plot.polygon
  is used by **panel.histogram()** and **panel.barchart()**,

box.3d
  is used by **panel.cloud()** and **panel.wireframe()**, and

regions, shade.colors
  are used by **panel.levelplot()** and **panel.wireframe()**.

Other settings are more general purpose. For example,

`superpose.symbol, superpose.line, superpose.polygon`
    are used for grouped displays in various contexts, whereas

`reference.line, add.line, add.text`
    are meant for secondary elements in a display, and are used in helper panel
    functions such as `panel.grid()` and `panel.text()`.

The `show.settings()` function produces a graphical display summarizing a
theme, as seen in Figure 7.4. A color version, summarizing the default color
theme, is also shown in the color plates. Further details can usually be inferred
from the setting names and the online documentation, and are not discussed
here.

7.2.1 Nonstandard settings

Some settings do not fall into the pattern described above and deserve a
separate discussion.

`clip`
    This parameter controls clipping separately for panels and strips. The
    default is

$$\text{clip = list(panel = "on", strip = "on")}$$

    that is, graphical output produced by the panel and strip functions will
    be clipped to the extent of the panel and strip regions.

`fontsize`
    This parameter controls the baseline font size (before `cex` is applied) for
    all text and points in the plot.

`grid.pars`
    This parameter is initially unset, but can be used to specify global defaults
    for parameters of the underlying **grid** engine that cannot be otherwise spec-
    ified. Examples include `lex` and `lineend`. A full list can be found on the
    `?gpar` help page in the **grid** package.

`layout.heights, layout.widths`
    These parameters control the amount of vertical and horizontal space allo-
    cated for the rows and columns that make up the layout of a lattice display.
    At the time of writing, every page of a lattice plot has rows allocated for
    (from top to bottom)
    1. A `main` label
    2. A legend (`key`)
    3. A common axis at the top (for `relation = "same"`)

4. One or more `strips` (one for each row of panels)
5. One or more `panels`
6. Axes at the bottom of each panel (e.g., for `relation = "free"`)
7. Spaces for the `between` argument
8. A common axis at the bottom (for `relation = "same"`)
9. An `xlab` below the panel(s)
10. A `key` at the bottom
11. A sub-title

Of course, not all these components are used in every plot. The layout also includes rows that are just for spaces (padding) between components. All these rows have a default height, and the `layout.heights` parameter can be used to specify multipliers for the default. The exact names and their current settings can be obtained by

```
> str(trellis.par.get("layout.heights"))
List of 18
 $ top.padding       : num 1
 $ main              : num 1
 $ main.key.padding  : num 1
 $ key.top           : num 1
 $ key.axis.padding  : num 1
 $ axis.top          : num 1
 $ strip             : num 1
 $ panel             : num 1
 $ axis.panel        : num 1
 $ between           : num 1
 $ axis.bottom       : num 1
 $ axis.xlab.padding : num 1
 $ xlab              : num 1
 $ xlab.key.padding  : num 1
 $ key.bottom        : num 1
 $ key.sub.padding   : num 1
 $ sub               : num 1
 $ bottom.padding    : num 1
```

For example, all the components with a name ending in "**padding**" can be set to 0 to make the layout as little wasteful of screen real estate as possible, while still allocating the minimum amount required for labels, legends, and the like. Some components, such as `panel`, `strip`, and `between`, are replicated for a display with multiple rows, and these components can be specified as a vector to achieve interesting results. The `layout.widths` parameter similarly controls the widths of columns in the layout; we leave the details for the reader to figure out.

`axis.components`

This parameter can be used to control the amount of space allocated for axis tick marks. It is rarely useful in practice.

7.3 Non-graphical options

A second set of settings is also maintained by lattice; these can be queried and modified using the functions `lattice.getOption()` and `lattice.options()`, which are analogous in behavior to `getOption()` and `options()`. These settings are global (not device-specific) and typically not graphical in nature, and primarily intended as a developer tool that allows experimentation with minimal code change. Because of its limited usefulness to the casual user, we do not discuss the available options extensively; the interested reader can find out more by inspecting the result of `lattice.options()` and reading the corresponding help page.

7.3.1 Argument defaults

One use of `lattice.options()` that is worth mentioning is in determining global defaults. The default layout in a lattice display counts rows from the bottom up, as in a graph, and not from the top down, as in a table. This is contrary to what many users expect. This behavior can be easily altered by specifying `as.table = TRUE` in a high-level call, but one might prefer to set a global preference instead by changing the default. This can be achieved with

```
> lattice.options(default.args = list(as.table = TRUE))
```

Default values can be set in this manner for several high-level arguments, including `aspect`, `between`, `page`, and `strip`. Some arguments in other functions also derive their defaults from settable options. The most useful of these is the `theme` argument of `trellis.device()`, which obtains its default from `lattice.getOption("default.theme")`.

7.4 Making customizations persistent

Customized themes can be made to persist across sessions using the R startup mechanism (see `?Startup`). There are two ways to do this, depending on whether lattice is automatically loaded during startup. In either case, the idea is to specify a default for the `theme` argument of `trellis.device()` through the options mechanism. Other options can be set at the same time. If lattice is to be loaded on startup, the following code might be included in `.First()` to change the default of `as.table` to TRUE and to make the standard color theme the default for all devices.

```
lattice.options(default.args = list(as.table = TRUE))
lattice.options(lattice.theme = standard.theme("pdf"))
```

A more sophisticated approach is to set a hook function that will be called only when lattice is attached through a call to `library()` or `require()`.

```
setHook(packageEvent("lattice", "attach"),
        function(...) {
            lattice.options(default.args = list(as.table = TRUE))
            lattice.options(default.theme =
                function() {
                    switch(EXPR = .Device,
                           postscript = ,
                           pdf = standard.theme(color = FALSE),
                           standard.theme("pdf", color = TRUE))
                })
        })
```

In this case the default theme is a function rather than a list, which uses the standard black and white theme as default for the **pdf()** and **postscript()** devices, and the standard color theme for all others.

8

Plot Coordinates and Axis Annotation

In this chapter, we discuss how the coordinate system for each panel is determined, how axes are annotated, and how one might control these in a lattice display. Control is possible at several levels, with a trade-off between the degree of control desired and the amount of effort required to achieve it.

8.1 Packets and the prepanel function

The controls discussed in this chapter can be broadly classified into two groups. Those in the first group relate to the determination of the coordinate system and axis limits for the panels (the rectangular regions within which graphics are drawn). Those in the second are concerned with how this coordinate system is described in the plot, typically through the use of tick marks and labels outside the panels. A grasp of the process determining the panel limits is essential to understand both sets of controls.

As described in Chapter 2, each combination of levels of the conditioning variables defining a *"trellis"* object gives rise to a *packet*. Loosely speaking, a packet is the data subset that goes into a panel representing a particular combination. Not all packets in a *"trellis"* object need end up in a plot of the object, and some may be repeated; however, a panel's limits are always determined by the entire collection of packets, and two panels with the same packet will have identical limits. In other words, limits are a property of packets, not panels. The rules determining these limits are described next.

Each panel area is a rectangular region in the Euclidean plane, and is defined completely by a horizontal and a vertical interval.[1] Even when the data being plotted are not intrinsically numeric (e.g., a categorical variable such as

---

[1] This is technically true even for functions such as cloud() and splom(), which are clearly different from other high-level functions, and bypass the controls described here. For these functions, scales and axis annotation are effectively controlled by the corresponding panel functions.

variety of oats), low-level plotting routines used to create the display require a numeric coordinate system. We refer to this as the *native coordinate system* of a panel. Given a packet, the prepanel function is responsible for determining a minimal rectangle in the native coordinate system that is sufficient to display that packet. It is implicitly assumed that any larger rectangle will also be sufficient for this purpose. Note that the minimal rectangle may depend not only on the packet but also on how it will eventually be displayed; for example, the maximum height of a histogram will differ greatly depending on whether it is a frequency, density, or relative frequency histogram, and to a lesser extent on the choice of bins.

Each high-level function comes with a default rule determining this minimal rectangle, and the `prepanel` argument gives the user further control. The rules governing the use of this argument are somewhat involved, and although the details are important, they are not immediately relevant. For the moment, assume that we have a rule to determine a minimal rectangle for each packet. A fuller discussion of the `prepanel` argument is postponed until later in this chapter.

8.2 The `scales` argument

8.2.1 Relation

There are three alternative schemes that prescribe, depending on how the panels are to relate to each other in the Trellis display, how the set of minimal rectangles collectively determines the final rectangles for each packet. The most common situation is to require all panels to have the same rectangle. This is achieved by choosing that common rectangle to be the one that minimally encloses all the individual rectangles. The second option is to allow completely independent rectangles, in which case the minimal rectangles are retained unchanged for each packet. The third option is to allow separate rectangles for each packet, but require their widths and heights (in the respective native coordinate systems) to be the same, with the intent of making differences comparable across panels, even if absolute positions are not (see Figure 10.6 for an example). In this case, each rectangle is expanded to make it as wide as the widest and as tall as the tallest rectangles. These rules can be selected by specifying `scales = "same"`, `scales = "free"`, and `scales = "sliced"`, respectively, in any high-level lattice call.

The description above is an oversimplification because in practice we often want to specify the relation between panels separately for the horizontal and vertical axes. This too can be achieved through the `scales` argument; for example,

```
scales = list(x = "same", y = "free")
```

leads to a common horizontal range and independent vertical ranges. More generally, the `scales` argument can also be used to specify a variety of other control parameters. In its general form, `scales` can be a list containing components called x and y, affecting the horizontal and vertical axes, each of which in turn can be lists containing parameters in `name = value` form. Parameters can also be specified directly as components of `scales`, in which case they affect both axes. For parameters specified both in `scales` and the x or y components, the values in the latter are given precedence.

As illustrated above, both `scales` and its x and y components can be a character string specifying the rule used to determine the packet rectangles. In the presence of other control parameters, this is no longer possible, and the string needs to be specified as the `relation` component. Thus, `scales = "free"` is equivalent to `scales = list(relation = "free")`, and

```
scales = list(x = "same", y = "sliced")
```

is equivalent to

```
scales = list(x = list(relation = "same"),
              y = list(relation = "sliced"))
```

Two other possible components of `scales` are involved in determining the panel coordinates, namely, `limits` and `axs`. It is difficult to discuss the purpose of these controls without first describing the prepanel function. For this reason, their discussion is likewise postponed until later in this chapter. Most other components of `scales` affect the drawing of tick marks and labels to annotate the axes. These are described next.

8.2.2 Axis annotation: Ticks and labels

Axis annotation is ultimately performed by the so-called *axis function*, which defaults to `axis.default()`, but can be overridden by the user. Other important functions under user control are ones that automatically determine tick mark locations and labels when these are not explicitly specified by the user. These functions, described later in this chapter, allow detailed control over axis annotation. However, such control is usually unnecessary, because some degree of control is already provided by components of the `scales` argument. We now list these components, noting that they apply only as long as the default axis annotation functions are used.

log

> This parameter controls whether the data will be log-transformed. It can be a scalar logical, and defaults to `FALSE`, in which case no transformation is applied. If `log = TRUE`, the data are log-transformed with base 10. Other bases can be specified using a numeric value such as `log = 2`. The natural logarithm can be specified by `log = "e"`. The choice of base does not alter the panel display, but can affect the location and ease of interpretation of the tick marks and labels. The `log` component is ignored with a warning in certain situations (e.g., for factors).

A non-default value of `log` has two effects. First, the relevant primary variable is suitably transformed. This happens before it is passed to the prepanel and panel functions, which are in fact never aware of this transformation.[2] Second, this affects how the default axis labels are determined. Specifically, pretty tick mark locations are chosen in the transformed scale, but the labels nominally represent values in the original scale by taking the form `base^at`,[3] where `at` represents tick mark locations in the transformed scale.

draw

This parameter must be a scalar logical (`TRUE` or `FALSE`). If it is `FALSE`, the axes are not drawn at all, and the parameters described below have no effect.

alternating

This parameter is applicable only if `relation = "same"`. In that case, axes are not drawn separately for all panels, but only along the "boundary" of the layout. In other words, axes are drawn only along the bottom (respectively, top) of panels in the bottom- (top-) most row and the left (right) of panels in the left- (right-) most column.[4] Axis annotation can consist of tick marks and accompanying labels. The tick marks are always drawn (unless suppressed by other parameters), but labels can be omitted in a systematic manner using the `alternating` parameter. Specifically, `alternating` can be a numeric vector, each of whose elements can be 0, 1, 2, or 3. When it applies as a parameter in the `x` (respectively, `y`) component of `scales`, it is replicated to be as long as the number of columns (rows) in the layout. The values are interpreted as follows: for a row with value 0, labels are not drawn on either side (left or right); for value 1, labels are only drawn on the left; for value 2, labels are only drawn on the right; and finally, for value 3, labels are drawn on both sides. Similarly, for columns, values of 1 and 2 lead to labels on the bottom and top, 3 to labels on both sides, and 0 to labels on neither.

`alternating` can also be a logical scalar. `alternating = TRUE` is equivalent to `alternating = c(1, 2)` and `alternating = FALSE` to `alternating = 1`. This explains the name of the parameter; `alternating = TRUE` causes labels to alternate between bottom (left) and top (right) in successive columns (rows). This is the default for numeric axes, where it

---

[2] For example, `panel.lmline()` will fit a linear regression to the transformed values, which may not be what one expects.

[3] This is clearly not the best solution, but determining nice tick mark locations on a logarithmic scale is a difficult problem. See later sections for examples of alternatives.

[4] As a special case, axes are also drawn on the right of the last panel on the page, even if it is not in the rightmost column.

helps avoid overlapping labels in adjacent panels.

tick.number

> This parameter acts as a suggested number of tick marks. Whether it will be used at all depends on the nature of the relevant variable; for example, it is honored for numeric (continuous) axes, but ignored for factors and shingles, because there is no reasonable basis for the selective omission of some labels in those cases.

at

> The automatic choice of tick mark locations can be overridden using the **at** parameter. When **relation = "same"**, **at** should be a numeric vector specifying the tick mark locations, or **NULL**, which is equivalent to **at = numeric(0)**. When **relation = "free"** or **"sliced"**, **at** can still be a numeric vector (in which case it is used for all panels), but can also be a list. This list should be exactly as long as the number of packets. Each element can be a numeric vector or **NULL**. Alternatively, it could also be logical (both **TRUE** and **FALSE** are acceptable), in which case that particular packet falls back to the default choice of **at**.

> The numeric locations of the tick marks must be specified in the native coordinates of the panel. This is true whether or not the axis is numeric. For factors and shingles, the ith level is usually encoded by the value i.

labels

> By default, labels are chosen automatically to correspond to the **at** values. This default choice can be overridden using the **labels** parameter. Like **at**, it can be a vector, or a list when **relation** is not **"same"**. Labels can be character strings as well as "expressions", allowing LaTeX-like mathematical annotation (see **?plotmath**). If a component is logical, the default rule is used to determine labels for that packet. If the lengths of corresponding components of **at** and **labels** do not match, the result is undefined.

abbreviate

> This is a logical flag, indicating whether the labels should be abbreviated using the **abbreviate()** function. Thic can be useful for long labels (e.g., for factors), especially on the x-axis.

minlength

> This is passed on to the **abbreviate()** function if **abbreviate = TRUE**.

format

> This is used as the **format** for *"POSIXct"* variables. See **?strptime** for a

description of valid values.

tck

This parameter controls the length of tick marks, and is interpreted as a numeric multiplier for the default length. If tck = 0, ticks are not drawn at all. Negative values cause tick marks to face inwards, which is generally a bad idea, but is sometimes desired as a matter of style. tck can be a vector of length 2, in which case the first element affects the left (respectively, bottom) axis and the second affects the right (top) axis.

rot

This parameter can be used to specify an angle (in degrees) by which the axis labels are to be rotated. It can be a vector of length 2, to control left and right (bottom and top) axes separately.

font, fontface, fontfamily

These parameters specify the font for axis labels.

cex, col, alpha

These parameters control other characteristics of the axis labels. cex is a numeric multiplier to control character sizes. Like rot, it can be a vector of length 2, to control left and right (bottom and top) axes separately. col controls color and alpha controls partial transparency on some devices.

col.line, alpha.line, lty, lwd

These parameters control graphical characteristics of the tick marks. Note that col.line does not affect the color of panel boundaries, which may lead to unexpected results. However, parameters for tick marks and panel boundaries both default to the "axis.line" settings (see Chapter 7 for details), whereas parameters for labels default to the "axis.text" settings. Together, this gives explicit control over each component individually.

8.2.3 Defaults

The defaults for the components of scales may be different for different high-level functions. This is achieved through a special argument called default.scales which the casual user should not be concerned about except to realize the role it plays in determining the defaults. Any parameter specified in default.scales serves as the default value of the corresponding parameter in scales.[5] For the more common parameters, the global defaults (used when no value is specified in either scales or default.scales) are

---

[5] One important point to note is that parameters specific to a particular axis in default.scales can only be overridden by a similarly specific parameter in scales.

$$\text{relation} = \text{"same"}$$
$$\log = \text{FALSE}$$
$$\text{draw} = \text{TRUE}$$
$$\text{alternating} = \text{TRUE}$$
$$\text{tick.number} = 5$$
$$\text{abbreviate} = \text{FALSE}$$
$$\text{minlength} = 4$$
$$\text{tck} = 1$$
$$\text{format} = \text{NULL}$$

Most other parameters are graphical parameters that default to the settings active during plotting. One special case is `rot`, which defaults to 0 if `relation = "same"`, but for other values of `relation`, it defaults to 0 for the x component and 90 for the y component.

`default.scales` is used primarily in situations where one of the axes generally represents a categorical variable (factor or shingle). For such axes, the defaults change to

$$\text{tck} = 0$$
$$\text{alternating} = \text{FALSE}$$
$$\text{rot} = 0$$

so that tick marks are omitted, the location of the labels do not alternate (saving space if the labels are long), and the labels are not rotated even when `relation` is not `"same"`. In `splom()`, `draw` defaults to `FALSE`, and much of the functionality of `scales` is accomplished instead by the `pscales` argument of `panel.pairs()`.

8.2.4 Three-dimensional displays: `cloud()` and `wireframe()`

The normal interpretation of "horizontal" and "vertical" axes makes no sense in the `cloud()` and `wireframe()` functions. There, the `scales` argument instead controls how the bounding box is annotated. As before, components can be specified directly, or in the x, y, or z components for specific axes. Many of the same parameters apply in this case, whereas many do not (and several should, but currently have no effect). There are two new parameters.

`arrows`

This parameter controls whether the annotation will be in the form of tick marks and labels, or just as an arrow encoding the direction of the axis. An arrow is used if `arrows = TRUE` (the default), and tick marks and labels are drawn otherwise. Both can be suppressed with `draw = FALSE`.

distance

Labels describing the axes (xlab, ylab, and zlab) are drawn along edges of the bounding box. This parameter controls how far these labels are from the box. distance should be a scalar if it is specified in axis-specific components, but if specified as a component of scales directly, it should be (and is recycled if not) a vector of length 3, specifying distances for the x, y, and z labels.

8.3 Limits and aspect ratio

8.3.1 The prepanel function revisited

As briefly mentioned earlier, the prepanel function is responsible for determining a minimal rectangle big enough to contain the graphical encoding of a given packet. Because this graphic is produced by the panel function, the prepanel function has to be chosen in concordance with it, and may in principle require all the information available to the panel function. For this reason, the prepanel function is usually called with exactly the same arguments[6] as the panel function, once for every packet. An important distinction is that the prepanel function is called as part of the process creating the *"trellis"* object, whereas the panel function is only called during plotting.

The return value of the prepanel function determines the minimal bounding rectangle for a packet, but it can also affect the axis labels and aspect ratio. In full generality, the return value can be a list consisting of components xlim, ylim, xat, yat, dx, and dy. Each high-level function has a default rule to calculate these quantities, so a user-specified prepanel function is not required to return all of these components; any missing component will be replaced by the corresponding default. The interpretation and effect of these components are described below.

xlim, ylim

These components together define a minimal bounding rectangle for the graphic to be created by the panel function given the same data packet. Two general forms are acceptable: a numeric vector of length 2 (as returned by range() for numeric data), and a character vector of arbitrary length (as returned by levels() for a factor). The first form is typical for numeric and date–time data, and xlim (or ylim, as the case may be) is interpreted as the horizontal (vertical) range of the rectangle. The second form is typical for factors, and is interpreted as a range containing c(1, length(xlim)), with the character vector determining labels at tick positions $1, 2, \ldots, $ length(xlim). If no other explicit specification of limits is made (e.g., through the high-level arguments xlim and ylim, or the

---

[6] Actually, some arguments may be dropped if the function does not accept them, and the list may be different for prepanel and panel functions.

limits component of scales), then the actual limits of the panels are guaranteed to contain the limits returned by the prepanel function.

The prepanel function is responsible for providing a meaningful return value for these components when the data contain missing or infinite values, or when they are empty (zero length). When nothing else is appropriate, xlim and ylim should be NA.

The limits returned by the prepanel function are usually extended (or padded) by a certain amount (configurable through lattice.options()), to ensure that points on or near the boundary do not get clipped. This behavior may be suppressed by specifying axs = "i" as a component of scales. The default behavior corresponds to axs = "r".

xat, yat

When xlim (or ylim) is a character vector, this is taken to mean that the scale should include the first n integers, where n is the length of xlim (ylim). The elements of the character vector are used as the default labels for these n integers. Thus, to make this information consistent between panels, the xlim or ylim values should represent all the levels of the corresponding factor, even if some are not used within that particular panel. To make relation = "free" or relation = "sliced" behave sensibly in such cases, an additional component xat (yat) may be returned by the prepanel function, which should be a subset of 1:n, indicating which of the n values (levels) are actually represented in the panel.

dx, dy

The dx and dy components are numeric vectors of the same length, together defining slopes of line segments used for banking computations when aspect = "xy", as described below.

8.3.2 Explicit specification of limits

The axis limits computed through the above mechanism can be overridden using the xlim and ylim arguments in high-level lattice functions. These should not be confused with the xlim and ylim components in the return value of prepanel, although they serve the same purpose and have the same valid forms. Specifically, the xlim and ylim arguments can either be numeric vectors of length 2, specifying an interval, or a character vector of length n, in which case the numeric data range is taken to be the interval $[1, n]$ with a suitable padding. As with the at and labels parameters of scales, xlim and ylim can also be specified on a per-packet basis when relation = "free". In this case, they have to be lists, with each component a numeric or character vector as above, or NULL in which case the default limits are used for the corresponding packet. The value of xlim or ylim is ignored when the corresponding

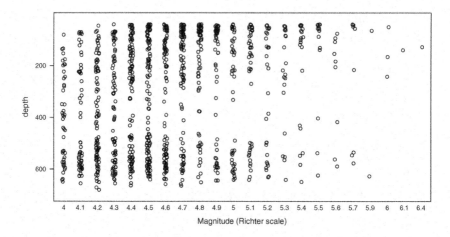

Figure 8.1. Rerendering of Figure 3.16, plotting the depth of earthquake epicenters against their magnitudes. The orientation of the y-axis has been reversed, so that depth now increases downwards.

`relation = "sliced"`. Alternatively, explicit limits can be specified as the `limits` components of `scales`. `scales$x$limits` is interpreted as `xlim` and `scales$y$limits` as `ylim`.

Although numeric limits usually take the form `c(minimum, maximum)`, this is not required. Limits in reverse order cause the corresponding axis to be reversed as well. For example, recall Figure 3.16, where `depth` was plotted on the vertical axis of a strip plot. Because depth is measured downwards from sea-level, it might be more natural to plot depth as increasing downwards in the graphic as well. This can be achieved with

```
> stripplot(depth ~ factor(mag), data = quakes, jitter.data = TRUE,
            ylim = c(690, 30),
            xlab = "Magnitude (Richter scale)")
```

An alternative that does not require knowing the data range beforehand is to make use of a custom prepanel function. Figure 8.1 is produced by

```
> stripplot(depth ~ factor(mag), data = quakes, jitter.data = TRUE,
            scales = list(y = "free", rot = 0),
            prepanel = function(x, y, ...) list(ylim = rev(range(y))),
            xlab = "Magnitude (Richter scale)")
```

Note that this will not work when `relation = "same"`, because in that case the process of combining limits from different packets makes the final limit sorted (even though there is only one packet in this example).

8.3.3 Choosing aspect ratio by banking

Statisticians are used to thinking in terms of invariance under scale and location changes, and often pay little attention to the unit of data being graphed. Although it usually makes no difference to the graphical encoding of the data, it is easy to see that the choice of units (including the choice of a base when taking logarithms) affects how the scales are annotated, indirectly affecting how easy it is to visually decode coordinates in the graphic. A less well-understood factor is the choice of physical units, that is, how much space (in centimeters, say) a plot will occupy on the display medium (computer monitor, paper, etc.). This is partly important because display media have finite physical resolution, and lines or points too close to each other will obscure patterns in the data. However, this problem is obvious when it occurs, and steps can be taken to rectify it. A more subtle feature is the ratio between physical units in the vertical (y) and horizontal (x) directions in a graphic, also known as the aspect ratio.

The importance of the aspect ratio has been noted by several authors (see Cleveland et al., 1988). In many situations, a satisfactory aspect ratio can only be found by trial and error. An exact aspect ratio, in the form of a numeric scalar, can be specified as the **aspect** argument. **aspect** can also be one of the character strings "fill", "iso", and "xy". When **aspect = "fill"**, panels expand to fill up all available space.[7] When **aspect = "iso"**, the aspect ratio is chosen so that the relationship between the physical and data scales (units per cm) is the same on both axes. When **aspect = "xy"**, the aspect ratio is chosen using the 45° banking algorithm described by Cleveland et al. (1988), based on their observation that judgments about small changes in slope are made most accurately when the slopes are close to 45°. The exact calculations are performed by the **banking()** function,[8] assuming that the relevant slopes in the plot are defined by the **dx** and **dy** components returned by the prepanel function.

The default banking computation to choose the aspect ratio is particularly useful with time-series data, where order is important and the slopes of line segments joining successive points are meaningful. The following example uses the **biocAccess** dataset from the latticeExtra package, which records the number of hourly access requests to the http://www.bioconductor.org Web site during the months of January through May of 2007. Figure 8.2 is produced by

```
> data(biocAccess, package = "latticeExtra")
> xyplot(counts/1000 ~ time | equal.count(as.numeric(time),
                                  9, overlap = 0.1),
          biocAccess, type = "l", aspect = "xy", strip = FALSE,
```

---

[7] If necessary, initial layout calculations assume **aspect = 1** in this case.

[8] Actually, the function that performs the banking calculations is user-settable, and is obtained as **lattice.getOption("banking")**. The default is **banking()**, but this can be overridden to implement more sophisticated approaches to banking.

```
        ylab = "Numer of accesses (thousands)", xlab = "",
        scales = list(x = list(relation = "sliced", axs = "i"),
                      y = list(alternating = FALSE)))
```

Apart from banking, this example also illustrates the use of a date–time object as a primary variable, which affects how the x-axis is annotated.

For unordered data, the dx and dy components computed by the default prepanel function are less useful, and alternative computations may be more appropriate in some situations. For example, one might want to bank based on the slopes of a smoothed version of the data. This can be done using a custom prepanel function that computes a suitable smooth and returns dx and dy values computed from the smoothed curve. Such a prepanel function is available in lattice for LOESS smoothing, and is called prepanel.loess(). It can be used, with the Earthquake dataset, to produce Figure 8.3 as follows.

```
> data(Earthquake, package = "MEMSS")
> xyplot(accel ~ distance, data = Earthquake,
         prepanel = prepanel.loess, aspect = "xy",
         type = c("p", "g", "smooth"),
         scales = list(log = 2),
         xlab = "Distance From Epicenter (km)",
         ylab = "Maximum Horizontal Acceleration (g)")
```

Two other predefined prepanel functions are available in lattice. These are prepanel.lmline(), which is similar to prepanel.loess(), but fits a simple linear regression model instead, and prepanel.qqmathline(), used with qqmath(), which fits a line through the first and third quartile pairs.

8.4 Scale components and the axis function

Although the scales argument can contain parameters affecting both, axis annotation is in principle distinct from the determination of panel coordinates. Low-level control over annotation, beyond what is possible with scales, is provided through two mechanisms. The first is a pair of functions, supplied as the xscale.components and yscale.components arguments, that compute tick mark positions and labels. The second is the axis function, specified as the axis argument, that actually renders the annotation. The axis function typically uses the results of xscale.components and yscale.components as well as scales, but is not required to do so. Using custom replacements for xscale.components or yscale.components while retaining the default axis function has the advantage that the right amounts of space for the tick marks and labels are automatically allocated.

8.4.1 Components

One situation where the default choice of tick marks and labels can clearly be improved is when using logarithmic scales. We continue with the Earthquake

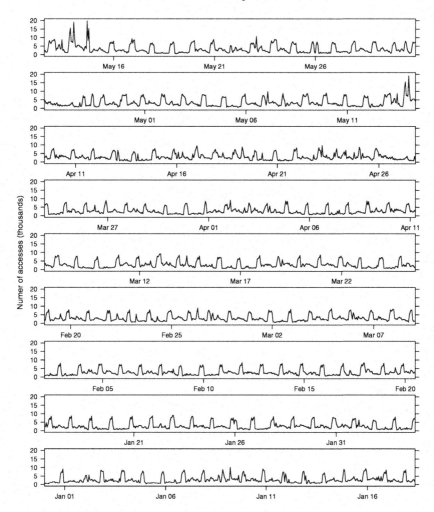

Figure 8.2. The hourly number of accesses to the `http://www.bioconductor.org` Web site during January through May of 2007. The aspect ratio has been chosen automatically using the 45° banking rule. The time axis has been split up into intervals to make use of the space available; such "cut-and-stack" plots are often useful with time-series data, and we encounter them again in Chapter 10. Figure 14.2 gives a more informative visualization of these data that makes effective use of some preliminary numerical analysis.

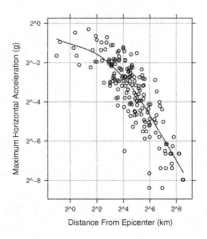

Figure 8.3. Rerendering of Figure 5.9, with automatically chosen aspect ratio. The predefined prepanel function **prepanel.loess()** is used, so that slopes of the LOESS smooth are used for the banking calculations. Note that the determination of the aspect ratio is unrelated to the display itself; if the **type** argument is omitted from the call, the display will not include the smooth, but the aspect ratio will remain unchanged.

example, with both x- and y-axes logarithmic, and try out some alternative ideas, implemented using the component functions. We only consider logarithms taken with base 2, but this can be adjusted as necessary.

Both **xscale.components** and **yscale.components** must return a list, with components **bottom** and **top** for the former, and components **left** and **right** for the latter, in addition to a component **num.limit** giving the numerical range of the limits as a vector of length 2. Details on the exact form of these components are not described here, but are available in the help page for **xscale.components.default()**. One interesting fact is that unlike **scales**, these allow the tick mark lengths to be vectorized, thus making major and minor tick marks possible.

In the following custom **yscale.components** function, we use the default components as a starting point, which allows us to bypass the uninteresting minutiae. We intend to have different labels on the two sides, so we next make the **right** component a copy of **left**. For both these components, the locations of the labels are kept unchanged, but the labels are modified. The labels on the left are turned into expressions, which leads to powers of 2 being rendered as superscripts. The labels on the right are converted into values in the original scale, possibly as fractions for negative powers, using the **fractions()** function from the MASS package. The final function is

```
> yscale.components.log2 <- function(...) {
      ans <- yscale.components.default(...)
```

```
ans$right <- ans$left
ans$left$labels$labels <-
    parse(text = ans$left$labels$labels)
ans$right$labels$labels <-
    MASS::fractions(2^(ans$right$labels$at))
ans
}
```

A more ambitious approach is to determine tick mark locations afresh in the original scale, ignoring the default computations. We can adapt the axTicks() function for this purpose, to define a new function called logTicks, which takes a numeric range lim, and returns locations within the range that take the form $i \times 10^j$, where i takes the values specified in the loc argument.

```
> logTicks <- function (lim, loc = c(1, 5)) {
    ii <- floor(log10(range(lim))) + c(-1, 2)
    main <- 10^(ii[1]:ii[2])
    r <- as.numeric(outer(loc, main, "*"))
    r[lim[1] <= r & r <= lim[2]]
}
```

This in turn can be used to define a custom xscale.components function:

```
> xscale.components.log2 <- function(lim, ...) {
    ans <- xscale.components.default(lim = lim, ...)
    tick.at <- logTicks(2^lim, loc = c(1, 3))
    ans$bottom$ticks$at <- log(tick.at, 2)
    ans$bottom$labels$at <- log(tick.at, 2)
    ans$bottom$labels$labels <- as.character(tick.at)
    ans
}
```

Note that it suffices to change the **bottom** component, as the **top** component takes the same value by default. Both these custom replacements are used below to produce Figure 8.4.

```
> xyplot(accel ~ distance | cut(Richter, c(4.9, 5.5, 6.5, 7.8)),
         data = Earthquake, type = c("p", "g"),
         scales = list(log = 2, y = list(alternating = 3)),
         xlab = "Distance From Epicenter (km)",
         ylab = "Maximum Horizontal Acceleration (g)",
         xscale.components = xscale.components.log2,
         yscale.components = yscale.components.log2)
```

As noted earlier, the component functions allow tick mark lengths to be vectorized, making it fairly easy to add minor tick marks. Figure 8.5 gives an example of this feature, using a variant of the custom components function used earlier.

```
> xscale.components.log10 <- function(lim, ...) {
    ans <- xscale.components.default(lim = lim, ...)
    tick.at <- logTicks(10^lim, loc = 1:9)
```

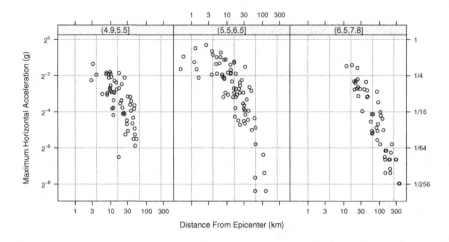

Figure 8.4. Fancy labels for logarithmic axes (compare with the axis annotation in Figure 8.3). The **alternating** parameter has been used to force axis labels on both the left and right sides simultaneously. The annotation is usually the same on both sides, but is different in this example where a user-supplied function has been used to compute the tick mark positions and labels.

```
    tick.at.major <- logTicks(10^lim, loc = 1)
    major <- tick.at %in% tick.at.major
    ans$bottom$ticks$at <- log(tick.at, 10)
    ans$bottom$ticks$tck <- ifelse(major, 1.5, 0.75)
    ans$bottom$labels$at <- log(tick.at, 10)
    ans$bottom$labels$labels <- as.character(tick.at)
    ans$bottom$labels$labels[!major] <- ""
    ans$bottom$labels$check.overlap <- FALSE
    ans
}
> xyplot(accel ~ distance, data = Earthquake,
        prepanel = prepanel.loess, aspect = "xy",
        type = c("p", "g"), scales = list(log = 10),
        xlab = "Distance From Epicenter (km)",
        ylab = "Maximum Horizontal Acceleration (g)",
        xscale.components = xscale.components.log10)
```

Notice that logarithms are taken with base 10 in this example; the only effect this has on the panel display is to change the location of the reference grid lines.

8.4.2 Axis

Changing the components is not always enough, and sometimes one may want to take full control of axis drawing. One situation where this might be useful

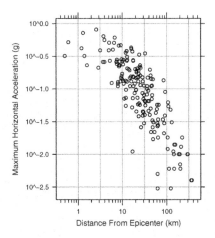

Figure 8.5. Another example of custom axis annotation: logarithmic axes with major and minor tick marks.

is when a single axis is used to represent multiple scales. Figure 8.6 plots a time-series of yearly temperatures in New Haven, CT, and annotates the temperature axis in both Celsius and Fahrenheit scales. This can be done using the following axis function, which uses `pretty()` to generate nice tick mark locations and `panel.axis()` (twice, in different colors) for the actual rendering. The same axis function must render the time axis as well, which our custom axis function handles simply by calling `axis.default()`.

```
> axis.CF <- function(side, ...) {
      if (side == "right") {
          F2C <- function(f) 5 * (f - 32) / 9
          C2F <- function(c) 32 + 9 * c / 5
          ylim <- current.panel.limits()$ylim
          prettyF <- pretty(ylim)
          prettyC <- pretty(F2C(ylim))
          panel.axis(side = side, outside = TRUE, at = prettyF,
                     tck = 5, line.col = "grey65", text.col = "grey35")
          panel.axis(side = side, outside = TRUE, at = C2F(prettyC),
                     labels = as.character(prettyC),
                     tck = 1, line.col = "black", text.col = "black")
      }
      else axis.default(side = side, ...)
  }
```

Figure 8.6 is produced by

```
> xyplot(nhtemp ~ time(nhtemp), aspect = "xy", type = "o",
         scales = list(y = list(alternating = 2, tck = c(1, 5))),
         axis = axis.CF, xlab = "Year", ylab = "Temperature",
```

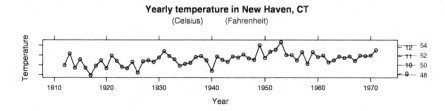

Figure 8.6. A custom axis function, providing calibration of temperature in both the Celsius and Fahrenheit scales. A legend has been added to describe the colors (see Chapter 9). Note the use of `tck` in `scales`. This affects the allocation of space for the tick marks and labels, which would otherwise need to be done manually.

```
     main = "Yearly temperature in New Haven, CT",
     key = list(text = list(c("(Celsius)", "(Fahrenheit)"),
               col = c("black", "grey35")), columns = 2))
```

One important point is that the axis function is called multiple times for each panel (once for each side), so careless use can easily lead to confusion. It should also be noted that the features discussed in the last section are fairly recent additions to lattice. Consequently, they are perhaps not as well thought out as the more traditional parts of the API, and some details may need to be changed in the future.

9

Labels and Legends

In this chapter, we discuss annotation of lattice displays by adding labels and legends. As usual, there are various levels of control available to the user, with corresponding differences in the amount of work involved. Most common needs for annotation are satisfied by various labels giving descriptive names for the variables and titles for the entire plot. Legends are usually needed to explain the correspondence between varying graphical parameters such as color, plotting character, and so on, and the quantitative information they represent.

9.1 Labels

Most high-level lattice functions allow four standard labels: `main`, `sub`, `xlab`, and `ylab`. Apart from their positions, they are treated identically for the most part.[1] They can be specified as a character string, as an expression (in which case they are interpreted as LaTeX-like markup, see `?plotmath`), or as a list.[2] In the first two cases, the string or expression is used as the label. The label can be a vector, in which case the components are evenly spread out (this allows row- or column-specific labels). In the third case, when `xlab`, `ylab`, and so on, are lists, the label can be specified as the `label` component. Other components, usually graphical parameters, but possibly ones controlling placement, are passed on to the grid function `textGrob()` to construct a suitable label. The `label` component can be omitted from the list, in which case the default label is used.

By default, `main` and `sub` are omitted in most displays, and `xlab` and `ylab` default to something appropriate, usually the expression for the corresponding variable in the formula, except when they are factors, in which case the label is omitted. Type

---

[1] `cloud()` and `wireframe()` interpret `xlab` and `ylab` differently, and allow a `zlab`.
[2] For more flexibility, they can also be specified as an arbitrary grob.

```
> demo("labels", package = "lattice")
```

to see some usage examples.

9.2 Legends

Legends, also called keys, usually serve to clarify the meaning of different graphical parameters (symbols, colors, etc.) used in a graphic. They are particularly important in grouped displays (where data from different groups are superposed within panels) and displays where a color gradient encodes a numeric variable (e.g., false-color level plots of three-dimensional surfaces). Legends can also be useful in other contexts; common examples are ones identifying orientation or scale in maps.

In some ways, legends are a weak point in the Trellis design. In the uses described above, as in most other uses, a legend describes features of the display created by the panel function. However, the Trellis model of separating the control of different elements of a display does not include any formal mechanism for direct communication between the processes controlling the panel display and the legend. Consequently, the only general approach that allows useful legends to be created automatically is to have both processes draw from a common source of information. For the collection of high-level functions built into the lattice package, this works reasonably well through the use of the auto.key and colorkey arguments. To understand these arguments though, we must first discuss the underlying processes that generate legends.

9.2.1 Legends as **grid** graphical objects

Although this fact is not overly emphasized in this book, the lattice package uses the low-level tools provided by the grid package to do all rendering. This choice is nowhere as important as it is in the context of legends. grid allows the creation of sophisticated "graphical objects" (grobs) that can not only be plotted, but also queried to determine their width and height. This is important in order to allocate the right amount of space for them, especially for legends, because they may have quite arbitrary structure. For full generality, legends in a lattice plot can be specified as arbitrary grobs. For most purposes, it suffices to use the predefined functions draw.key() and draw.colorkey(), which both produce specialized and highly structured grobs of a certain kind. As we soon show, the user needs no knowledge of grid or grobs to use these functions.

The draw.key() function

The draw.key() function accepts an argument called key and returns a grob.[3] The grob represents a legend containing a series of components laid out in the

---

[3] It can also draw the grob, a fact we use to create Figure 12.1.

form of a table, possibly divided into multiple blocks. The components can be text, points, lines, or rectangles. These can appear in an arbitrary order, and each component can be repeated or be completely absent. The legend can also have a title.

All this can be achieved through the `key` argument, which must be a list. All its components must be named, of which the names `text`, `points`, `lines`, and `rectangles` may be repeated. Each component named `text` contributes a column of text in the legend, each component named `points` contributes a column of points, and so on, in the order in which they appear in `key`. Each of these components must be lists, containing zero or more graphical parameters specified as vectors. The only special cases are the `text` components, which must have a vector of character strings or expressions as their first component.

Graphical parameters are usually specified as components of the `text`, `points`, `lines`, and `rectangles` lists. They can also be specified directly as components of `key`, but with lower precedence. Valid graphical components are `cex`, `col`, `lty`, `lwd`, `font`, `fontface`, `fontfamily`, `pch`, `adj`, `type`, `size`, `angle`, and `density`, although not all of these apply to all components. Most of these parameters are standard, with the following exceptions.

`adj`

> This parameter controls justification of text. Meaningful values are between 0 (left justified) and 1 (right justified).

`type`

> This parameter is only relevant for lines; `"l"` results in a plain line, `"p"` produces a point, and `"b"` and `"o"` produce both together.

`size`

> This parameter determines the width of rectangles and lines in character widths.

`angle, density`

> These parameters are included for compatibility with S-PLUS code, but are currently unimplemented. They are intended to control the details of cross-hatching in rectangles.

Unless otherwise specified (see `rep` below), it is assumed that all columns (except the `text` ones) will have the same number of rows. This common number is taken to be the largest of the lengths of the graphical components, including the ones specified directly in `key`. For a `text` component, the number of rows is the length of its first component, which must be a character or expression vector. Several other components of `key` affect the final legend, as described next.

`rep`

> This can be a scalar logical, defaulting to TRUE, in which case all non-text columns in the key are replicated to be as long as the longest. This can be suppressed by specifying `rep = FALSE`, in which case the length of each column will be determined by components of that column alone.

divide

> When `type` is `"b"` or `"o"` in a `lines` component, each line is divided by these many point symbols.

title

> A character string or expression giving a title for the key.

cex.title

> A `cex` factor for the title.

lines.title

> Amount of vertical space allocated for the title, in multiples of its own height. Defaults to 2.

transparent

> A scalar logical, indicating whether the key area should have a transparent background. Defaults to `FALSE`, but see the next entry.

background

> The background color for the legend, which defaults to the default background setting. Note that this default is often `"transparent"`, in which case `transparent = FALSE` will have no visible effect.

border

> This can either be a color for the border, or a scalar logical. In the latter case, the border color is black if `border = TRUE`, and no border is drawn if it is `FALSE` (the default).

between

> This can be a numeric vector giving the amount of blank space (in terms of character widths) surrounding each column. The specified width is split equally on both sides of a column.

padding.text

> This indicates how much space (padding) should be left above and below each row containing text, in multiples of the default. This padding is in addition to the normal height of any row that contains text, which is the minimum amount necessary to contain all the text entries.

columns

> The name of this parameter is somewhat misleading, because it specifies not the number of columns in the key, but rather the number of column-blocks into which the key is to be divided. Specifically, rows of the key are divided into these many blocks, which are then drawn side by side.

between.columns

> Space between column blocks, in addition to `between`.

The draw.colorkey() function

The `draw.colorkey()` function is in many ways much simpler. It too accepts an argument called `key`, and produces a grob that represents a color gradient, along with tick marks and labels that provide calibration for the colors. The legend is defined by the following components of `key`.

space

> The intended location of the key, possible values being `"left"`, `"right"` (the default), `"top"`, and `"bottom"`. This only affects the grob to the extent that it determines the orientation of the color gradient (vertical in the first two cases, horizontal in the last two) and the location of the tick marks relative to the gradient (always facing "outwards").

col

> The vector of colors used in the legend. The number of colors actually shown is one less than `length(at)` (see below); `col` is replicated if it is shorter, and a subset chosen by sampling linearly if longer. The same rule is used by `panel.levelplot()` and `panel.wireframe()` when appropriate.

at

> It is always assumed that the colors supplied represent discrete bins along some numeric interval (although the tick mark labels can be manipulated to suggest otherwise). `at` is a numeric vector defining these bins. Specifically, they determine where the colors change, and must be in ascending order. There is no requirement for the `at` values to be equispaced.

labels

> This can be a character vector (or expression) for labeling the `at` values, but this use is unusual. More commonly, `labels` is a list, which itself has one or more of the components `at`, `labels`, `cex`, `col`, `font`, `fontface`, and `fontfamily`. This works much as does `scales` (see Chapter 8), in the sense that the `at` and `labels` components, defining the tick mark locations and labels, are determined automatically if unspecified.

tick.number

> Suggested number of ticks, used when the tick mark locations are unspecified.

width

> A multiplier to control the width, or rather the thickness, of the key. When the key is horizontal (`space` is `"top"` or `"bottom"`), this actually controls the height.

height

> One interesting feature of the grobs produced by `draw.colorkey()` is that they are "expandable" in one direction; the color bar does not have an absolute length, but expands to fit in the space available. This component determines what proportion of the available space the legend will occupy. As with `width`, the name of this component is misleading when `space` is `"top"` or `"bottom"`.

9.2.2 The `colorkey` argument

A color gradient as produced by `draw.colorkey()` is only relevant for two high-level lattice functions: `levelplot()` and `wireframe()` (the latter only when `drape = TRUE`). For these functions, the legend can be controlled by the `colorkey` argument. The legend can be suppressed with `colorkey = FALSE`, and enabled with `colorkey = TRUE`, the latter being the default whenever a color gradient is used. Alternatively, `colorkey` can be a list, in which case it is used as the `key` argument in `draw.colorkey()`. The most common use of this is to change the location of the legend, for example, with `colorkey = list(space = "top")`. The only component of `key` without a reasonable default in `draw.colorkey()` is `at`, which in the case of `levelplot()` and `wireframe()` defaults to the corresponding `at` argument in the high-level function. Adding a color key in other high-level functions is possible, but more involved, as we have seen in Figure 5.6.

9.2.3 The `key` argument

Unlike `draw.colorkey()`, which is designed for a fairly specific purpose, `draw.key()` is intended to be quite general. The `key` argument, accepted by all high-level functions (including `levelplot()` and `wireframe()`), allows legends produced by `draw.key()` to be added to a plot. Such a `key` argument can be a list as accepted by `draw.key()`, with the following additional components also allowed.

`space`

> This specifies the intended location of the key, possible values being `"left"`, `"right"`, `"top"` (the default), and `"bottom"`.

`x, y, corner`

> These components specify an alternative positioning of the legend inside the plot region. `x` and `y` determine the location of the corner of the key given by `corner`. Common values of `corner` are `c(0, 0)`, `c(1, 0)`, `c(1,1)`, and `c(0,1)`, which denote the corners of the unit square, but fractional values are also allowed. `x` and `y` should be numbers between 0 and 1, giving coordinates with respect to either the whole display area, or just the subregion containing the panels. The choice is controlled by `lattice.getOption("legend.bbox")`, which can be `"full"` or `"panel"` (the default).

Figure 8.6 gives an example of a simple but nontrivial legend produced using the `key` argument, as does Figure 9.2 later in this chapter. These examples demonstrate the flexibility of `draw.key()`. However, in practice, most legends are associated with a grouping variable, supplied as the `groups` argument. The generality of `draw.key()` is unnecessary for such legends, which typically have exactly one column of text (containing the levels of `groups`), and at most one column each of points, lines, or rectangles. Furthermore, if the different graphical parameters associated with different levels of `groups` are obtained from the global settings, the contents of these columns are also entirely predictable.

One way to create such standard legends is to use the `Rows()` function, which is useful in extracting a subset of graphical parameters suitable for use as a component in `key`. Consider the `Car93` dataset, which contains information on several 1993 passenger car models (Lock, 1993; Venables and Ripley, 2002), and can be loaded using

```
> data(Cars93, package = "MASS")
```

For our first example, we plot the midrange price against engine size, conditioning on `AirBags`, with `Cylinders` as a grouping variable. To make things interesting, we leave out the level `"rotary"`, which is represented only once in the data:

```
> table(Cars93$Cylinders)

   3    4    5    6    8 rotary
   3   49    2   31    7      1
```

As the first five levels of `Cylinders` are plotted, we can extract the corresponding default graphical settings as

```
> sup.sym <- Rows(trellis.par.get("superpose.symbol"), 1:5)
> str(sup.sym)
List of 6
 $ alpha: num [1:5] 1 1 1 1 1
 $ cex  : num [1:5] 0.7 0.7 0.7 0.7 0.7
 $ col  : chr [1:5] "#000000" "#000000" "#000000" ...
 $ fill : chr [1:5] "#EBEBEB" "#DBDBDB" "#FAFAFA" ...
 $ font : num [1:5] 1 1 1 1 1
 $ pch  : num [1:5] 1 3 6 0 5
```

This can now be used in a call to `xyplot()` to produce Figure 9.1.

```
> xyplot(Price ~ EngineSize | reorder(AirBags, Price), data = Cars93,
         groups = Cylinders, subset = Cylinders != "rotary",
         scales = list(y = list(log = 2, tick.number = 3)),
         xlab = "Engine Size (liters)",
         ylab = "Average Price (1000 USD)",
         key = list(text = list(levels(Cars93$Cylinders)[1:5]),
                    points = sup.sym, space = "right"))
```

This computation can be simplified using a convenience function called `simpleKey()`, which returns a list suitable for use as the `key` argument. The first argument to `simpleKey()` (`text`) must be a vector of character strings or expressions, giving the labels in the text column. It can also be given logical arguments `points`, `lines`, and `rectangles` specifying whether a corresponding column will be included in the key; if `TRUE`, the graphical parameters for the corresponding component are constructed using calls to `trellis.par.get()` and `Rows()` as above. The settings `"superpose.symbol"` is used for `points`, `"superpose.line"` for `lines`, and `"superpose.polygon"` for `rectangles`. Further arguments to `simpleKey()` are simply retained as elements of the list returned. Thus, an alternative call producing Figure 9.1 is

```
> xyplot(Price ~ EngineSize | reorder(AirBags, Price), data = Cars93,
         groups = Cylinders, subset = Cylinders != "rotary",
```

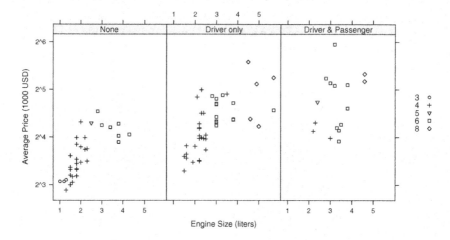

Figure 9.1. Average (of basic and premium) price of cars plotted against engine size. The data are separated into panels representing number of airbags (ordered by mean price), and the number of cylinders is used as a grouping variable within each panel.

```
scales = list(y = list(log = 2, tick.number = 3)),
xlab = "Engine Size (liters)",
ylab = "Average Price (1000 USD)",
key = simpleKey(text = levels(Cars93$Cylinders)[1:5],
                space = "right", points = TRUE))
```

9.2.4 The problem with settings, and the auto.key argument

This approach, although appearing to be effective at first glance, breaks down if we consider the possibility of changes in the settings. As we saw in Chapter 7, presentation of a graphic is not entirely defined by its contents; that is, the same *"trellis"* object can be plotted multiple times using different themes, resulting in the use of different graphical parameters. This is especially relevant for grouped displays, where color might be used to distinguish between groups when available, and other parameters such as plotting character and line type used otherwise. This choice is determined by the theme in use when the object is plotted, and thus, it is impossible to determine the legend prior to that point. The problem with the approach described above, using simpleKey(), is that the legend is instead determined fully when the object is created.

The solution is to postpone the call to simpleKey() until plotting time. This can be achieved through the auto.key argument, which can simply be a list containing arguments to be supplied to simpleKey(). Thus, yet another call that produces Figure 9.1 is

```
> xyplot(Price ~ EngineSize | reorder(AirBags, Price), data = Cars93,
         groups = Cylinders, subset = Cylinders != "rotary",
         scales = list(y = list(log = 2, tick.number = 3)),
         xlab = "Engine Size (liters)",
         ylab = "Average Price (1000 USD)",
         auto.key = list(text = levels(Cars93$Cylinders)[1:5],
                         space = "right", points = TRUE))
```

This version will update the legend suitably when the resulting object is plotted with different themes. In fact, the auto.key approach allows for more intelligent defaults, and it is usually possible to omit the text component (which defaults to the group levels) as well as the points, lines, and rectangles components (which have function-specific defaults). One can simply use auto.key = list(space = "right") in the above call, or even auto.key = TRUE which would use the default space = "top". Unfortunately, in both these cases, the omitted level ("rotary") will be included in the legend.

9.2.5 Dropping unused levels from groups

For conditioning variables and primary variables that are factors, levels that are unused after the application of the subset argument in a high-level call are usually omitted from the display. It is difficult to do the same with unused levels of groups. This is a consequence of the design; groups is passed to the panel function as a whole, and appropriate panel-specific subsets are extracted using the subscripts argument. subscripts refers to rows in the original data before applying subset, and so, groups must also be available in its entirety. Dropping levels inside the panel function is not an option, as some levels may be present in some panels, but not in others.

This behavior can sometimes be frustrating, and often the simplest solution is to subset the data beforehand, possibly using the subset() function. One more operation is required to omit the unused levels, as subset() does not do so itself. In the following call, which is yet another way to produce Figure 9.1, this is done inline when specifying groups.

```
> xyplot(Price ~ EngineSize | reorder(AirBags, Price),
         data = subset(Cars93, Cylinders != "rotary"),
         groups = Cylinders[, drop = TRUE],
         scales = list(y = list(log = 2, tick.number = 3)),
         xlab = "Engine Size (liters)",
         ylab = "Average Price (1000 USD)",
         auto.key = list(space = "right"))
```

Many other examples that use auto.key can be found throughout this book.

9.2.6 A more complicated example

Although rare, there are nonetheless occasions where auto.key is not sufficient. We finish this section with one such example, where two grouping

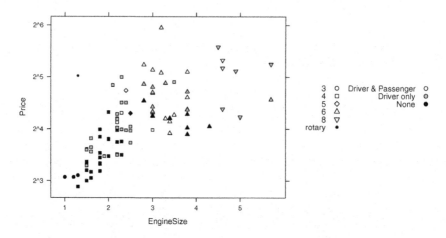

Figure 9.2. An alternative to Figure 9.1, with both **Cylinders** and **AirBags** now used as grouping variables, encoded by different graphical attributes (plotting character and fill color). The associated legend has to be constructed explicitly using the **key** argument.

variables are used concurrently, with levels distinguished by varying two different graphical parameters. In particular, our goal is to produce an alternative form of Figure 9.1, where in addition to **Cylinders**, **AirBags** is also a grouping variable rather than a conditioning variable. Consequently, the legend must contain two columns of text, one for each grouping variable, of different lengths. Figure 9.2 is produced by the following code.

```
> my.pch <- c(21:25, 20)
> my.fill <- c("transparent", "grey", "black")
> with(Cars93,
       xyplot(Price ~ EngineSize,
              scales = list(y = list(log = 2, tick.number = 3)),
              panel = function(x, y, ..., subscripts) {
                  pch <- my.pch[Cylinders[subscripts]]
                  fill <- my.fill[AirBags[subscripts]]
                  panel.xyplot(x, y, pch = pch,
                               fill = fill, col = "black")
              },
              key = list(space = "right", adj = 1,
                         text = list(levels(Cylinders)),
                         points = list(pch = my.pch),
                         text = list(levels(AirBags)),
                         points = list(pch = 21, fill = my.fill),
                         rep = FALSE)))
```

The use of with() allows us to refer to elements of Cars93 by name inside the panel function.

9.2.7 Further control: The legend argument

Legends produced by draw.key() can be quite general, but they are ultimately limited in scope. The legend argument, although more involved in its use, provides far greater flexibility. This flexibility is afforded by the ability to specify the legend as an arbitrary grob, or alternatively a function, called at plotting time, that produces a grob. We give an example illustrating the use of this feature, but do not discuss it in much detail as it is rarely useful to the casual user. Details can be found in the online documentation.

Our example is a heatmap, which is a graphical representation of a hierarchical clustering of rows and/or columns of a matrix. We consider again the USArrests dataset, which tabulates the number of arrests for various crimes in 1973 per 100,000 residents in the 50 U.S. states. Our goal is to cluster the states, which can be done with the hclust() function.

```
> hc1 <- hclust(dist(USArrests, method = "canberra"))
> hc1 <- as.dendrogram(hc1)
```

We coerce the result to a "dendrogram" object before manipulating it further. The next step is to find a permutation of the states that arranges them in the "right" order; there is more than one such permutation, and we determine one that retains grouping by region (given by the state.region dataset) as much as possible.

```
> ord.hc1 <- order.dendrogram(hc1)
> hc2 <- reorder(hc1, state.region[ord.hc1])
> ord.hc2 <- order.dendrogram(hc2)
```

We are now almost ready to draw our heatmap. Our first attempt might be

```
> levelplot(t(scale(USArrests))[, ord.hc2])
```

where the states are reordered, each variable is scaled to make the units comparable, and the data matrix is transposed to produce a tall (rather than wide) display. Of course, this will not show the actual clustering, which is where the legend argument comes in. The lattice package has no built-in support for plotting dendrograms, but it does allow new legends to be designed and used. The dendrogramGrob() function in the latticeExtra package conveniently produces a grob representing a given dendrogram, and can be used as follows to produce Figure 9.3.

```
> library("latticeExtra")
> region.colors <- trellis.par.get("superpose.polygon")$col
> levelplot(t(scale(USArrests))[, ord.hc2],
            scales = list(x = list(rot = 90)),
            colorkey = FALSE,
            legend =
            list(right =
```

```
list(fun = dendrogramGrob,
     args =
     list(x = hc2, ord = ord.hc2,
          side = "right", size = 10, size.add = 0.5,
          add = list(rect =
            list(col = "transparent",
                 fill = region.colors[state.region])),
          type = "rectangle"))))
```

Here, the normal color key is disabled as the units lose their meaning after scaling. Instead, we put in the dendrogram as the legend on the right side. The specification of legend is fairly simple in an abstract sense; it is a list with a component right indicating that the legend should be placed to the right of the panel(s), which in turn consists of components fun, which is a function that returns a grob, and args, which is a list of arguments supplied to fun. For the interpretation of the arguments provided to dendrogramGrob(), see the corresponding help page.

Writing a function such as dendrogramGrob() requires familiarity with the grid package.[4] For those interested in traveling that road, dendrogramGrob() can serve as a useful template.

9.3 Page annotation

Another form of annotation is available through the page argument to a high-level plot, which must be a function that is executed once for every page, with the page number as its only argument. The function must use grid-compliant plotting commands (which include lattice panel functions), and is called with the whole display area as the default viewport and the native coordinate system set to the unit square $[0, 1] \times [0, 1]$. An obvious use of this argument is to add page numbers to multipage lattice plots; for example, as

```
page = function(n) {
    panel.text(lab = sprintf("Page %d", n), x = 0.95, y = 0.05)
}
```

Such a function could be set as the global default:

```
> lattice.options(default.args = list(page = function(n) {
      panel.text(lab = sprintf("Page %d", n), x = 0.95, y = 0.05)
  }))
```

in which case all subsequent lattice plots would include a page number. Another possible use of page is to perform some interactive task after a page is drawn, such as placing a legend by clicking on a location in the display area; an example is shown in Figure 12.1.

---

[4] In particular, making sure that the legend "expands" to exactly fit the panel, even when the plot is resized, involves the concepts of frames and packing.

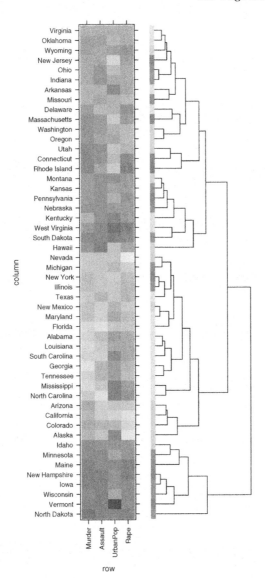

Figure 9.3. A heatmap created with the standard `levelplot()` function along with a nonstandard legend representing a hierarchical clustering. The thin strip at the root of the dendrogram represents a grouping of the states based on geographical location (south, northeast, etc.). Unlike the standard `heatmap()` function, this implementation puts no restrictions on the aspect ratio.

10

Data Manipulation and Related Topics

Now that we have had a chance to look at several types of lattice plots and ways to control their various elements, it is time to take another look at the big picture and introduce some new ideas. This chapter may be viewed as a continuation of Chapter 2; the topics covered are slightly more advanced, but generally apply to all lattice functions.

10.1 Nonstandard evaluation

Variables in the Trellis formula (along with those in groups and subset, if supplied) are evaluated in an optional data source specified as the data argument. This is usually a data frame, but could also be a list or environment. (Other types of data sources can be handled by writing new methods, as we see in Chapter 14.) When a term in the formula (or groups or subset) involves a variable not found in data, special scoping rules apply to determine its value; it is searched for in the environment of the formula, and if not found there, in the enclosing environment, and so on. In other words, the search does not start in the environment where the term is being evaluated, as one might expect. If no data argument is specified in a lattice call, the search for all variables starts in the environment of the formula. This behavior is similar to that in other formula-based modeling functions (e.g., lm(), glm(), etc.), and is commonly referred to as "standard nonstandard evaluation".

This is not an entirely academic issue. There are situations where this nonstandard scoping behavior is desirable, and others where it gives "unexpected" results. To get a sense of the issues involved, consider the following example where we revisit the choice of an optimal Box–Cox transformation for the gcsescore variable in the Chem97 data (Figure 3.7). Instead of choosing the transformation analytically, we might simply try out several choices and visualize the results. We might do this by first defining a function implementing the Box–Cox transformation

```
> boxcox.trans <- function(x, lambda) {
      if (lambda == 0) log(x) else (x^lambda - 1) / lambda
  }
```

which is then used to create a multipage PDF file.

```
> data(Chem97, package = "mlmRev")
> trellis.device(pdf, file = "Chem97BoxCox.pdf",
                 width = 8, height = 6)
> for (p in seq(0, 3, by = 0.5)) {
      plot(qqmath(~boxcox.trans(gcsescore, p) | gender, data = Chem97,
                  groups = score, f.value = ppoints(100),
                  main = as.expression(substitute(lambda == v,
                                                  list(v = p)))))
  }
> dev.off()
```

In this example, the variable p in the formula is not visible in the **data** argument, and according to the nonstandard evaluation rules, it is searched for (and found) in the environment in which the formula was defined. This is the right thing to do in this case; we would not have wanted to use any other variable named p that might have been visible in the environment where the terms in the formula are actually evaluated. On the other hand, someone used to the standard lexical scoping behavior in R might think that the following is a reasonable alternative.

```
> form <- ~ boxcox.trans(gcsescore, p) | gender
> qqboxcox <- function(lambda) {
      for (p in lambda)
          plot(qqmath(form, data = Chem97,
                      groups = score, f.value = ppoints(100),
                      main = as.expression(substitute(lambda == v,
                                                      list(v = p)))))
  }
> qqboxcox(lambda = seq(0, 3, by = 0.5))
```

However, this will either fail because p is not found, or worse, use the wrong value of p. Of course, this is a rather artificial and perhaps not very convincing example. Most of the real problems due to nonstandard evaluation arise when trying to implement new wrapper functions with similar semantics, especially because the nonstandard evaluation rules also apply to the **groups** and **subset** arguments. One standard solution is outlined in the final example in Chapter 14.

10.2 The extended formula interface

We have already encountered the **Titanic** dataset in Chapter 2. To use the data in a lattice plot, it is convenient to coerce them into a data frame, as we do here for the subset of adults:

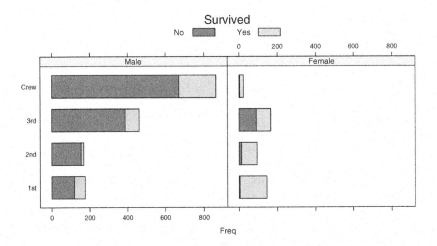

Figure 10.1. A bar chart showing the fate of adult passengers of the Titanic.

```
> Titanic1 <- as.data.frame(as.table(Titanic[, , "Adult" ,]))
> Titanic1
    Class    Sex Survived Freq
1    1st   Male       No  118
2    2nd   Male       No  154
3    3rd   Male       No  387
4   Crew   Male       No  670
5    1st Female       No    4
6    2nd Female       No   13
7    3rd Female       No   89
8   Crew Female       No    3
9    1st   Male      Yes   57
10   2nd   Male      Yes   14
11   3rd   Male      Yes   75
12  Crew   Male      Yes  192
13   1st Female      Yes  140
14   2nd Female      Yes   80
15   3rd Female      Yes   76
16  Crew Female      Yes   20
```

This form of the data is perfectly suited to our purposes. For example, Figure 10.1 can be produced from it by

```
> barchart(Class ~ Freq | Sex, Titanic1,
           groups = Survived, stack = TRUE,
           auto.key = list(title = "Survived", columns = 2))
```

Unfortunately, data may not always be as conveniently formatted. For example, these data could easily have been specified in the so-called "wide" format (as opposed to the "long" format above):

```
> Titanic2
  Class    Sex Dead Alive
1   1st   Male  118    57
2   2nd   Male  154    14
3   3rd   Male  387    75
4  Crew   Male  670   192
5   1st Female    4   140
6   2nd Female   13    80
7   3rd Female   89    76
8  Crew Female    3    20
```

This format is particularly common for longitudinal data, where multiple observations (e.g., over time) on a single experimental unit are often presented in one row rather than splitting them up over several rows (in which case covariates associated with the experimental units would have to be repeated). The formula interface described in Chapter 2 is not up to handling the seemingly simple task of reproducing Figure 10.1 from the data in the wide format.

The traditional solution is to transform the data into the long format before plotting. This can be accomplished using the **reshape()** function; in fact, our artificial example was created using

```
> Titanic2 <-
      reshape(Titanic1, direction = "wide", v.names = "Freq",
              idvar = c("Class", "Sex"), timevar = "Survived")
> names(Titanic2) <- c("Class", "Sex", "Dead", "Alive")
```

Unfortunately, **reshape()** is not the simplest function to use, and as this kind of usage is common enough, lattice provides a way to avoid calling **reshape()** by encoding the desired transformation within the formula. In particular, the part of the formula specifying primary variables (to the left of the conditioning symbol) can contain multiple terms separated by a + symbol, in which case they are treated as columns in the wide format that are to be concatenated to form a single column in the long format. Figure 10.2 is produced by

```
> barchart(Class ~ Dead + Alive | Sex, Titanic2, stack = TRUE,
           auto.key = list(columns = 2))
```

Notice that the new factor implicitly created (to indicate from which column in the wide format a row in the long format came) has been used for grouping without any explicit specification of the **groups** argument. This is the default behavior for high-level functions in which grouped displays make sense. One may instead want to use the new factor as a conditioning variable; this can be done by specifying **outer = TRUE** in the call. In either case, the **subscripts** argument, if used in a custom panel function, refers to the implicitly reshaped data.

An alternative interpretation of such formulae that avoids the concept of reshaping is as follows: the formula y1 + y2 ~ x | a should be taken to mean that the user wishes to plot both y1 ~ x | a and y2 ~ x | a, with **outer** determining whether the plots are to use the same or separate panels.

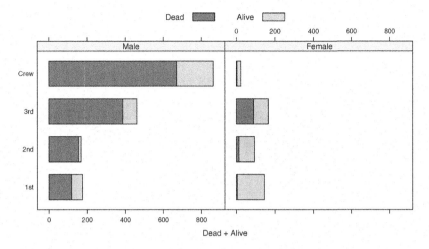

Figure 10.2. An alternative formulation of Figure 10.1, using data in the wide format. The plots are identical, except for the legend and the x-axis label.

This behavior is distributive, in the sense that the formula y1 + y2 ~ x1 + x2 will cause all four combinations (y1 ~ x1, y1 ~ x2, y2 ~ x1, and y2 ~ x2) to be displayed. To interpret y1 + y2 as a sum in the formula, one can use I(y1 + y2), which suppresses the special interpretation of +.

For another example where the use of the extended formula interface arises naturally, consider the Gcsemv dataset (Rasbash et al., 2000) in the mlmRev package.

```
> data(Gcsemv, package = "mlmRev")
```

The data record the GCSE exam scores of 1905 students in England on a science subject. The scores for two components are recorded: written paper and course work. The scores are paired, so it is natural to consider a scatter plot of the written and coursework scores conditioning on gender. Figure 10.3 is produced by

```
> xyplot(written ~ course | gender, data = Gcsemv,
          type = c("g", "p", "smooth"),
          xlab = "Coursework score", ylab = "Written exam score",
          panel = function(x, y, ...) {
              panel.xyplot(x, y, ...)
              panel.rug(x = x[is.na(y)], y = y[is.na(x)])
          })
```

where we use the predefined panel function panel.rug()[1] to encode the scores for cases where one component is missing (these would otherwise have been omitted from the plot). This plot clearly suggests an association between the

---

[1] See Chapter 13 for a full list of predefined panel functions.

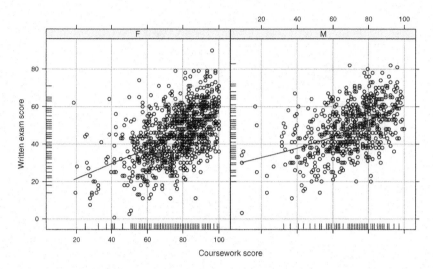

Figure 10.3. A scatter plot of coursework and written exam scores on a science subject, conditioned on gender. Scores missing in one variable are indicated by "rugs".

two scores. In the next plot, we ignore the pairing of the scores and look at their marginal distributions using a Q–Q plot. Figure 10.4 is produced by

```
> qqmath(~ written + course, Gcsemv, type = c("p", "g"),
          outer = TRUE, groups = gender, auto.key = list(columns = 2),
          f.value = ppoints(200), ylab = "Score")
```

This plot emphasizes two facts about the marginal distributions: boys tend to do slightly better than girls in the written exam whereas the opposite is true for coursework, and although the written scores fit a normal distribution almost perfectly, the coursework scores do not. In fact, the distributions of coursework scores have a positive probability mass at 100 (one might say that the "true" scores have been right censored), more so for girls than boys. Neither of these facts are unexpected, but they are not as obvious in the previous scatter plot.

10.3 Combining data sources with make.groups()

By itself, the formula interface is not flexible enough for all situations, and one often needs to manipulate the data before using them in a lattice call. One common scenario is when datasets of different lengths need to be combined. All terms in the Trellis formula (even in the extended form) should have the same length after evaluation. This restriction is naturally enforced when data is a data frame, but there is no explicit check for other data sources. This can sometimes be an issue if one is careless, especially with "univariate" formu-

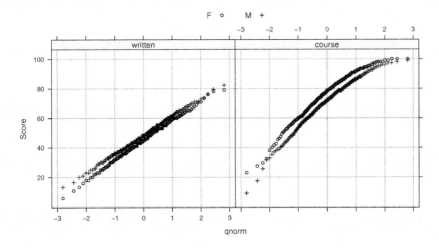

Figure 10.4. Normal quantile plots of written exam and coursework scores, grouped by gender. The distributions of written exam scores are close to normal, with males doing slightly better. The distributions of coursework scores are skewed, with several full scores for females (note that only a subset of quantiles has been plotted), who do considerably better. The comparison is visually more striking when color is used to distinguish between the groups.

lae as used in Q–Q plots and histograms. To make this point, consider this somewhat artificial example: Among continuous probability distributions, the exponential distribution is unique in having the property (often referred to as the memoryless property) that left truncation is equivalent to an additive shift in the induced distribution. To demonstrate this using a Q–Q plot, we generate truncated and untruncated observations from the standard exponential distribution.

```
> x1 <- rexp(2000)
> x1 <- x1[x1 > 1]
> x2 <- rexp(1000)
```

In view of the preceding discussion, one might be tempted to try something along the lines of

```
> qqmath(~ x1 + x2, distribution = qexp)
```

to create a grouped theoretical Q–Q plot, but this produces the wrong output because x1 and x2 are not of equal length. The correct approach is to combine the vectors and form a suitable grouping variable, as in

```
> qqmath( ~ c(x1, x2), distribution = qexp,
          groups = rep(c("x1", "x2"), c(length(x1), length(x2))))
```

This is of course tedious, even more so when there are more objects to combine. A utility function designed for such situations is `make.groups()`, which

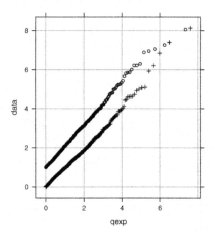

Figure 10.5. Theoretical quantile plot comparing the untruncated and truncated (at 1) exponential distributions. The plot illustrates the use of `make.groups()` as discussed in the text.

combines several vectors, possibly of different lengths, into a single data frame with two columns: one (`data`) concatenating all its arguments, the other (`which`) indicating from which vector an observation came. For example, we have

```
> str(make.groups(x1, x2))
'data.frame':   1772 obs. of  2 variables:
 $ data : num   2.31 2.35 2.49 1.51 ...
 $ which: Factor w/ 2 levels "x1","x2": 1 1 1 1 1 1 1 1 ...
```

We can thus produce Figure 10.5 with

```
> qqmath(~ data, make.groups(x1, x2), groups = which,
         distribution = qexp, aspect = "iso", type = c("p", "g"))
```

Multiple data frames with differing number of rows can also be combined using `make.groups()`, provided they have conformable columns. As an example, consider the `beavers` dataset (Reynolds, 1994; Venables and Ripley, 2002), which actually consists of two data frames `beaver1` and `beaver2`, recording body temperature of two beavers in north-central Wisconsin every ten minutes over a period of several hours.

```
> str(beaver1)
'data.frame':   114 obs. of  4 variables:
 $ day  : num   346 346 346 346 346 346 346 346 ...
 $ time : num   840 850 900 910 920 930 940 950 ...
 $ temp : num   36.3 36.3 36.4 36.4 ...
 $ activ: num   0 0 0 0 0 0 0 0 ...
> str(beaver2)
```

```
'data.frame':    100 obs. of  4 variables:
 $ day  : num  307 307 307 307 307 307 307 307 ...
 $ time : num  930 940 950 1000 1010 1020 1030 1040 ...
 $ temp : num  36.6 36.7 36.9 37.1 ...
 $ activ: num  0 0 0 0 0 0 0 0 ...
```

We can combine these in a single data frame using

```
> beavers <- make.groups(beaver1, beaver2)
> str(beavers)
'data.frame':    214 obs. of  5 variables:
 $ day  : num  346 346 346 346 346 346 346 346 ...
 $ time : num  840 850 900 910 920 930 940 950 ...
 $ temp : num  36.3 36.3 36.4 36.4 ...
 $ activ: num  0 0 0 0 0 0 0 0 ...
 $ which: Factor w/ 2 levels "beaver1","beaver2": 1 1 1 1 1 1 1 1 ...
```

The time of each observation is recorded in a somewhat nonstandard manner. To use them in a plot, one option is to convert them into hours past an arbitrary baseline using

```
> beavers$hour <-
      with(beavers, time %/% 100 + 24*(day - 307) + (time %% 100)/60)
```

The range of this new variable is very different for the two beavers (the two sets of measurements were taken more than a month apart), so plotting them on a common axis does not make sense. We could of course measure hours from different baselines for each beaver, but another alternative is to allow different limits using

```
> xyplot(temp ~ hour | which, data = beavers, groups = activ,
         auto.key = list(text = c("inactive", "active"), columns = 2),
         xlab = "Time (hours)", ylab = "Body Temperature (C)",
         scales = list(x = list(relation = "sliced")))
```

The result is shown in Figure 10.6. This is a natural use of "sliced" scales, as we want differences in time to be comparable across panels, even though the absolute values have no meaningful interpretation.

10.4 Subsetting

As with other formula-based interfaces in R (such as lm() and glm()), one can supply a subset argument to choose a subset of rows to use in the display. If specified, it should be an expression, possibly involving variables in data, that evaluates to a logical vector. The result should have the same length as the number of rows in the data, and is recycled if not. For example, the graphic in Figure 10.1 (which uses only the subset of adults) could have been obtained directly from the full Titanic data by

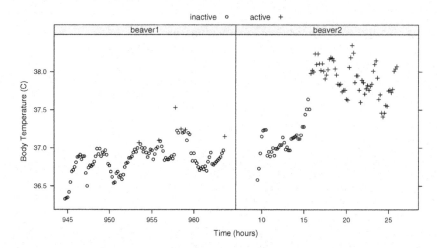

Figure 10.6. Body temperature of two beavers (over time) in north-central Wisconsin. The plotting symbols indicate periods of outside activity, which clearly affects body temperature.

```
> barchart(Class ~ Freq | Sex, as.data.frame(Titanic),
           subset = (Age == "Adult"), groups = Survived, stack = TRUE,
           auto.key = list(title = "Survived", columns = 2))
```

Subsetting becomes more important for larger datasets. To illustrate this, let us consider the USAge.df dataset in the latticeExtra package, which records estimated population[2] of the United States by age and sex for the years 1900 through 1979.

```
> data(USAge.df, package = "latticeExtra")
> head(USAge.df)
  Age  Sex Year Population
1   0 Male 1900      0.919
2   1 Male 1900      0.928
3   2 Male 1900      0.932
4   3 Male 1900      0.932
5   4 Male 1900      0.928
6   5 Male 1900      0.921
```

Figure 10.7 plots the population distribution for every tenth year starting with 1905:

```
> xyplot(Population ~ Age | factor(Year), USAge.df,
         groups = Sex, type = c("l", "g"),
         auto.key = list(points = FALSE, lines = TRUE, columns = 2),
         aspect = "xy", ylab = "Population (millions)",
         subset = Year %in% seq(1905, 1975, by = 10))
```

---

[2] Source: U.S. Census Bureau, http://www.census.gov.

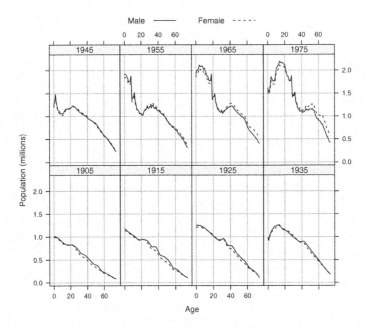

Figure 10.7. U.S. population distribution by gender, every ten years from 1905 through 1975.

The "baby boom" phenomenon of the late 1940s and 1950s is clearly apparent from the plot. It is clearer in the next representation, where each panel represents a specific age, and plots the population for that age over the years. Figure 10.8 is produced by

```
> xyplot(Population ~ Year | factor(Age), USAge.df,
         groups = Sex, type = "l", strip = FALSE, strip.left = TRUE,
         layout = c(1, 3), ylab = "Population (millions)",
         auto.key = list(lines = TRUE, points = FALSE, columns = 2),
         subset = Age %in% c(0, 10, 20))
```

In particular, the panel for age 0 represents the number of births (ignoring immigration, which is less important here than in the older age groups). A closer look at the panel for 20-year-olds shows an intriguing dip in the male population around 1918. To investigate this further, the next plot follows the population distribution by cohort; Figure 10.9 conditions on the year of birth, and is produced by

```
> xyplot(Population ~ Year | factor(Year - Age), USAge.df,
         groups = Sex, subset = (Year - Age) %in% 1894:1905,
         type = c("g", "l"), ylab = "Population (millions)",
         auto.key = list(lines = TRUE, points = FALSE, columns = 2))
```

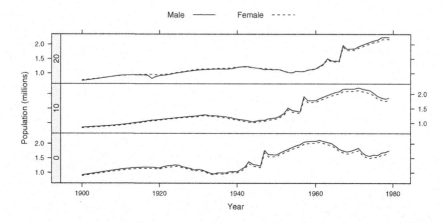

Figure 10.8. U.S. population by age over the years. The bottom panel gives the (approximate) number of births each year, clearly showing the baby boom after World War II. The top panel, which gives the population of 20-year-olds, shows a temporary drop in the male population around 1918.

Unlike in the previous plots, no individual is counted in more than one panel. The signs of some major event in or around 1918 is clear, and a closer look suggests that its impact on population varies by age and sex. The most natural explanation is that the fluctuation is related to the United States joining World War I in 1917; however, there is no similar fluctuation for World War II. As it turns out, armed forces stationed overseas were excluded from the population estimates for the years 1900-1939, but not for subsequent years.

10.4.1 Dropping of factor levels

A subtle point that is particularly relevant when using the `subset` argument is the rule governing dropping of levels in factors. By default, levels that are unused (i.e., have no corresponding data points) after subsetting are omitted from the plot. This behavior can be changed by the `drop.unused.levels` argument separately for conditioning variables and panel variables. The default behavior is usually reasonable, but the ability to override it is often helpful in obtaining a more useful layout. Note, however, that levels of a grouping variable are never dropped automatically. This is because unlike variables in the formula, subsetting of `groups` is done inside panel functions, and dropping levels in this case may inadvertently lead to inconsistency across panels or meaningless legends. A possible workaround is described in Chapter 9 in the context of the `auto.key` argument.

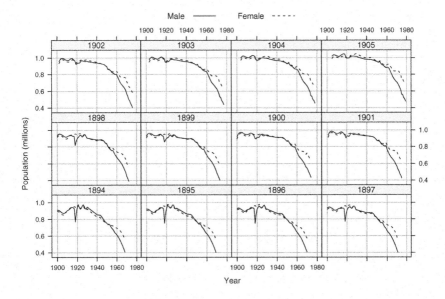

Figure 10.9. U.S. population by cohort (birth year), showing the effect of age and sex on the temporary drop in population in 1918. The use of broken lines for females leads to unequal visual emphasis on the two groups that is completely artificial; compare with the color version of this figure, which is included among the color plates.

10.5 Shingles and related utilities

We have briefly encountered shingles previously in Chapter 2. In this section, we take a closer look at the facilities available to construct and manipulate shingles. As an example, we use the `quakes` dataset, and look at how the number of stations reporting an earthquake is related to its magnitude. Figure 10.10 is produced by

```
> xyplot(stations ~ mag, quakes, jitter.x = TRUE,
         type = c("p", "smooth"),
         xlab = "Magnitude (Richter)",
         ylab = "Number of stations reporting")
```

Subject to certain assumptions, we might expect the counts to have a Poisson distribution, with mean related to earthquake magnitude. The documentation of the dataset notes that there are no quakes with magnitude less than 4.0 on the Richter scale, but a closer look at Figure 10.10 also reveals that there are none with less than ten reporting stations. This truncation makes it harder to decide whether the the expected count is a linear function of the magnitude, although the shape of the LOESS smooth for magnitude greater than 4.5 supports that conjecture. Another implication of the Poisson model is that the variance of the counts increases with the mean. We use shingles to investigate

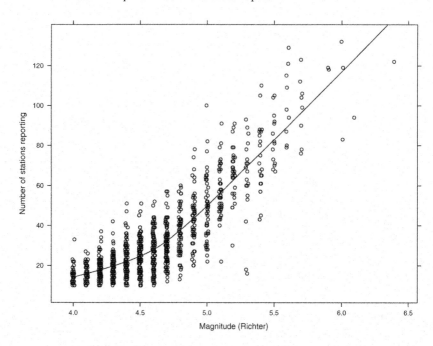

Figure 10.10. Number of stations recording earthquakes of various magnitudes. Earthquakes with less than ten stations reporting are clearly omitted from the data, which possibly explains the curvature in the LOESS smooth.

whether this seems to be true.[3] First, we construct a shingle from the numeric `mag` variable using the `equal.count()` function.

```
> quakes$Mag <- equal.count(quakes$mag, number = 10, overlap = 0.2)
```

This creates a shingle with ten levels, each represented by a numeric interval. The endpoints of the intervals are determined automatically (based on the data) so that roughly the same number of observations falls in each. The `overlap` argument determines the fraction of overlap between successive levels; in this case, 20% of the data in each interval should also belong to the next. The `overlap` can be negative, in which case there will be gaps in the coverage. It is also possible to create shingles with the intervals explicitly specified, using the `shingle()` function, as we soon show. The resulting levels and the corresponding frequencies can be inspected by summarizing the shingle:

```
> summary(quakes$Mag)
Intervals:
    min  max count
1  3.95 4.25   191
2  4.05 4.35   230
```

---

[3] This is by no means the only way to do so.

```
3   4.25 4.45    186
4   4.35 4.55    208
5   4.45 4.65    208
6   4.55 4.75    199
7   4.65 4.85    163
8   4.75 5.05    166
9   4.85 5.25    173
10 5.05 6.45     151
```

```
Overlap between adjacent intervals:
[1] 145   85 101 107 101   98   65 101   72
```

The nominal goals of having an equal number of observations in each level and an overlap of 20% between successive intervals have not quite been met, but this is a consequence of heavy rounding of the magnitudes, with only 22 unique values. A character representation of the levels, useful for annotation, is produced by

```
> as.character(levels(quakes$Mag))
 [1] "[ 3.95, 4.25 ]" "[ 4.05, 4.35 ]" "[ 4.25, 4.45 ]"
 [4] "[ 4.35, 4.55 ]" "[ 4.45, 4.65 ]" "[ 4.55, 4.75 ]"
 [7] "[ 4.65, 4.85 ]" "[ 4.75, 5.05 ]" "[ 4.85, 5.25 ]"
[10] "[ 5.05, 6.45 ]"
```

A visual representation of the shingle can be produced using the plot() method[4] for shingles. This actually creates a *"trellis"* object, which we store in the variable ps.mag for now instead of plotting it.

```
> ps.mag <- plot(quakes$Mag, ylab = "Level",
                 xlab = "Magnitude (Richter)")
```

Next, we create another *"trellis"* object representing a box-and-whisker plot with stations on the y-axis and the newly created shingle on the x-axis.

```
> bwp.quakes <-
      bwplot(stations ~ Mag, quakes, xlab = "Magnitude",
             ylab = "Number of stations reporting")
```

Finally, we plot these *"trellis"* objects together to produce Figure 10.11.

```
> plot(bwp.quakes, position = c(0, 0, 1, 0.65))
> plot(ps.mag, position = c(0, 0.65, 1, 1), newpage = FALSE)
```

Without the plot of the shingle, we would not be able to associate a level of the shingle to the numeric interval it represents. An alternative is to manually annotate the shingle levels, as in Figure 10.12, which is produced by

```
> bwplot(sqrt(stations) ~ Mag, quakes,
         scales =
         list(x = list(limits = as.character(levels(quakes$Mag)),
                       rot = 60)),
```

---

[4] See ?plot.shingle for details.

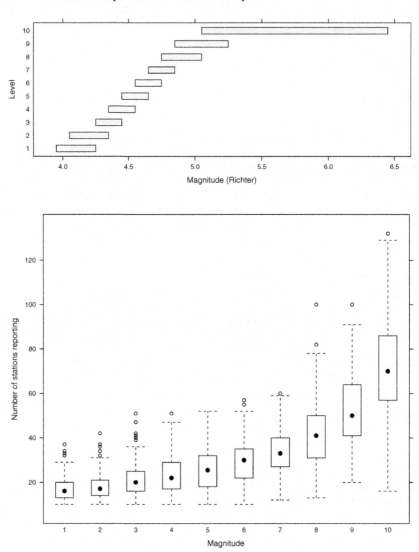

Figure 10.11. A box-and-whisker plot of the number of stations recording earth-quakes, with a shingle representing the earthquake magnitudes. A plot of the shingle at the top gives the association between levels of the shingle and the corresponding ranges of magnitude on the Richter scale.

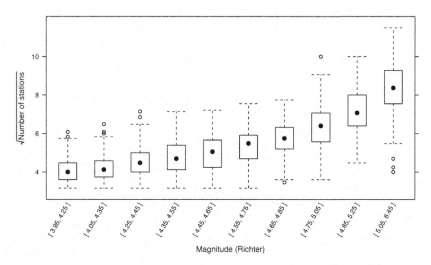

Figure 10.12. A variant of Figure 10.11. The *y*-axis now plots the square root of the number of stations reporting, and the range of each level of the shingle is given in the form of axis labels.

```
xlab = "Magnitude (Richter)",
ylab = expression(sqrt("Number of stations")))
```

where we additionally plot the number of reporting stations on a square root scale. The square root transformation is a standard variance stabilizing transformation for the Poisson distribution (Bartlett, 1936), and does seem to work reasonably well, given the omission of quakes reported by less than ten stations.

It is more common to use shingles as conditioning variables. In that case, the interval defining a level of the shingle is indicated relative to its full range by shading the strip region. If the exact numeric values of the interval are desired, one can add a plot of the shingle as in Figure 10.11. Another option is to print the numeric range inside each strip using a suitable strip function, as in the following call which produces Figure 10.13.

```
> qqmath(~ sqrt(stations) | Mag, quakes,
        type = c("p", "g"), pch = ".", cex = 3,
        prepanel = prepanel.qqmathline, aspect = "xy",
        strip = strip.custom(strip.levels = TRUE,
                            strip.names = FALSE),
        xlab = "Standard normal quantiles",
        ylab = expression(sqrt("Number of stations")))
```

The strip.custom() function used in this example is explained later in this chapter.

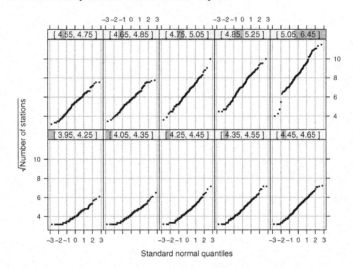

Figure 10.13. A normal quantile plot of the (square root of the) number of stations reporting earthquakes. The conditioning variable is a shingle, with numerical range shown in the strips. The truncation in the number of stations can be seen in the first few panels.

10.5.1 Coercion to factors and shingles

There are certain situations where lattice expects to find a "categorical variable" (i.e., either a factor or a shingle). This most obviously applies to conditioning variables; numeric variables are coerced to be shingles, and character variables are turned into factors. This coercion rule also applies in `bwplot()`, as well as a few other high-level functions, such as `stripplot()` and `barchart()`, which expect one of the axes to represent a categorical variable. If numeric variables are given for both axes, the choice of which one to coerce depends on the value of the `horizontal` argument.

This behavior can be frustrating, because there are occasions where we want a numeric variable to be used as a categorical variable in the display, yet retain the numeric scale for spacing and axis annotation. For example, we might want a plot with the same structure as Figure 10.10, but with the jittered point clouds for each value of `mag` replaced by a box-and-whisker plot. Unfortunately, attempting to do so with `bwplot()` will fail; the display produced by

```
> xyplot(stations ~ mag, quakes,
         panel = panel.bwplot, horizontal = FALSE)
```

will represent `mag` as a shingle with unique values equally spaced along the *x*-axis, and we will lose information about gaps in their values. One solution is to create a factor or shingle explicitly with empty levels. A simpler option

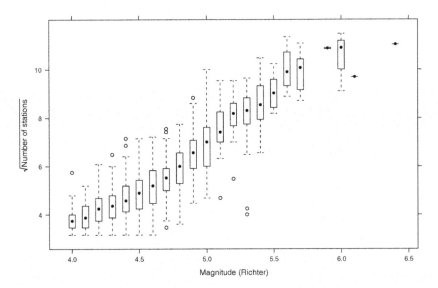

Figure 10.14. Box-and-whisker plots of number of reporting stations by magnitude of earthquakes. This time, the x variable is not a shingle, but the actual magnitude.

is not to use bwplot() at all, but instead use xyplot(), and borrow the panel function panel.bwplot(). Figure 10.14 is produced by

```
> xyplot(sqrt(stations) ~ mag, quakes, cex = 0.6,
         panel = panel.bwplot, horizontal = FALSE, box.ratio = 0.05,
         xlab = "Magnitude (Richter)",
         ylab = expression(sqrt("Number of stations")))
```

We need to tweak the box.ratio argument to account for the fact that successive boxes are not as far away from each other as panel.bwplot() expects.

10.5.2 Using shingles for axis breaks

Although shingles are commonly used for conditioning, they can be put to other interesting uses as well. In particular, a numeric variable can be used both in its original form (as a primary variable) and as a shingle (as a conditioning variable) in conjunction with the relation specification to create (possibly data-driven) scale breaks. As an example, consider the population density in the 50 U.S. states, based on population estimates from 1975:

```
> state.density <-
      data.frame(name = state.name,
                 area = state.x77[, "Area"],
                 population = state.x77[, "Population"],
                 region = state.region)
> state.density$density <- with(state.density, population / area)
```

We can produce a Cleveland dot plot of the raw densities using

```
> dotplot(reorder(name, density) ~ density, state.density,
          xlab = "Population Density (thousands per square mile)")
```

producing Figure 10.15. The plot is dominated by a few states with very high density, making it difficult to assess the variability among the remaining states. This kind of problem is usually alleviated by taking a logarithmic transformation, as we do later in Figures 10.19 through 10.21. However, another option is to create a break in the x-axis. There is achieved by creating a shingle, using the `shingle()` constructor, with suitable intervals specified explicitly.

```
> state.density$Density <-
      shingle(state.density$density,
              intervals = rbind(c(0, 0.2),
                                c(0.2, 1)))
```

This shingle can now be used as a conditioning variable to separate the states into two panels. Figure 10.16 is created using

```
> dotplot(reorder(name, density) ~ density | Density, state.density,
          strip = FALSE, layout = c(2, 1), levels.fos = 1:50,
          scales = list(x = "free"), between = list(x = 0.5),
          xlab = "Population Density (thousands per square mile)",
          par.settings = list(layout.widths = list(panel = c(2, 1))))
```

where the x-axis is allowed to be different for the two panels, and additionally the panel with more states is further emphasized by making it wider.

10.5.3 Cut-and-stack plots

Another use of shingles that is similar in spirit is to create so-called "cut-and-stack" plots (Cleveland, 1993). Time-series data are often best viewed with a low aspect ratio because local features are usually of more interest than overall trends. A suitable aspect ratio can usually be determined automatically using the 45° banking rule (`aspect = "xy"`), but this generally results in a short wide plot that does not make use of available vertical space. An easy way to remedy this is to divide up (cut) the time range into several smaller parts and stack them on top of each other. Shingles are ideal for defining the cuts because they allow overlaps, providing explicit continuity across panels. In fact, if we use the `equal.count()` function to create the shingle, all we need to specify is the number of cuts and the amount of overlap, as we did in Figure 8.2. We can wrap this procedure in a simple function:

```
> cutAndStack <-
      function(x, number = 6, overlap = 0.1, type = "l",
               xlab = "Time", ylab = deparse(substitute(x)), ...) {
      time <- if (is.ts(x)) time(x) else seq_along(x)
      Time <- equal.count(as.numeric(time),
                          number = number, overlap = overlap)
```

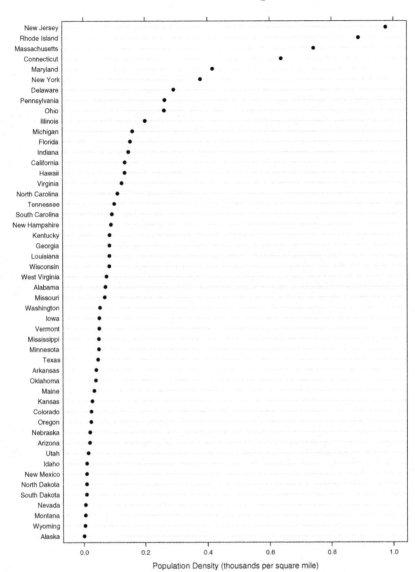

Figure 10.15. Estimated population density in U.S. states, 1975. A few extreme values dominate the plot.

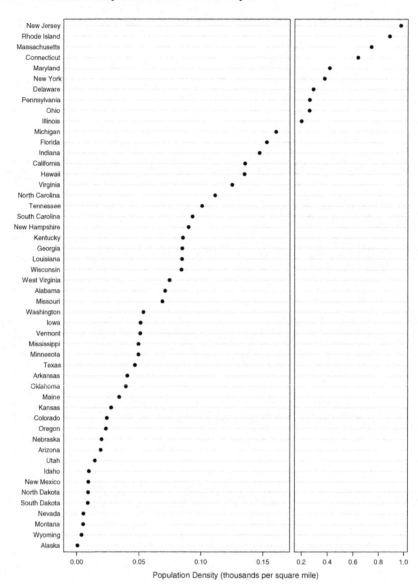

Figure 10.16. Estimated population density in U.S. states, with a break in the x-axis.

```
xyplot(as.numeric(x) ~ time | Time,
       type = type, xlab = xlab, ylab = ylab,
       default.scales = list(x = list(relation = "free"),
                             y = list(relation = "free")),
       ...)
}
```

We can then use this to create a cut-and-stack plot of a time-series object (class "*ts*") or any other numeric vector. Figure 10.17 is produced by

```
> cutAndStack(EuStockMarkets[, "DAX"], aspect = "xy",
              scales = list(x = list(draw = FALSE),
                            y = list(rot = 0)))
```

An alternative approach is presented in Chapter 14.

10.6 Ordering levels of categorical variables

Unlike ordinal categorical variables and numeric variables (and by extension shingles), levels of nominal variables have no intrinsic order. An extremely important, but rarely appreciated fact is that the visual order of these levels in a display has considerable impact on how we perceive the information encoded in the display. By default, when R creates factors from character strings, it defines the levels in alphabetical order (this can be changed using the `levels` argument to the `factor()` constructor), and this order is retained in lattice plots. In most cases, reordering the levels based on the data, rather than keeping the original arbitrary order, leads to more informative plots.

This fact plays a subtle but important role in the well-known barley dot plot shown in Figure 2.6. In this plot, the levels of variety, site, and year were all ordered so that the median yield within level went from lowest to highest. We reproduce this plot in Figure 10.18 alongside a slightly different version where the levels of site and variety are ordered alphabetically. Although the switch in direction at Morris, the primary message from the plot, is clear in both plots, the one on the right makes it easier to convince ourselves that the likely culprit is simply a mislabeling of the year for that one site, and no more elaborate explanation is required.

Reordering levels of a factor with respect to the values of another variable is most easily done using the `reorder()` function.[5] We have already used `reorder()` in Chapter 4 to produce Figure 4.7, and earlier in this chapter when looking at dot plots of population densities in the United States. Continuing with the latter example, we can create a dot plot with log densities on the *x*-axis using

---

[5] `reorder()` is a generic function, and documentation for the method we are interested in can be accessed using `?reorder.factor`.

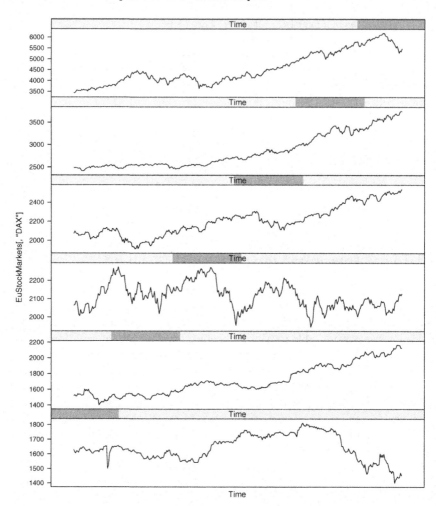

Figure 10.17. A cut-and-stack plot of stock market time-series data. The x variable is also used as a conditioning variable, with the effect of splicing up the x-axis across different panels. The y-axes are determined independently for each panel, emphasizing local variation rather than overall trends.

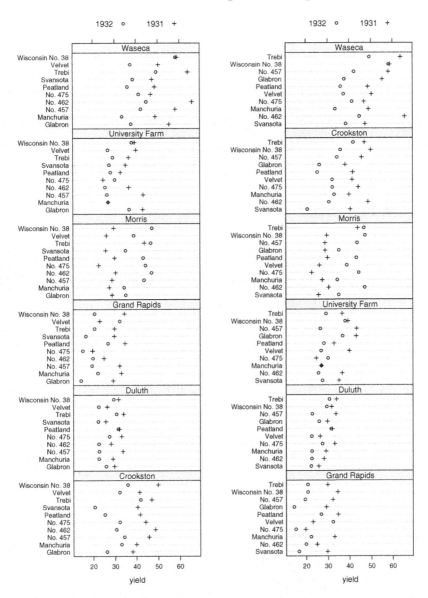

Figure 10.18. A grouped dot plot of the barley data with default (alphabetical by levels) and median (Figure 2.6) ordering. The ordered version makes it easier to convince ourselves that switching the year labels for Morris is enough to "fix" the data.

```
> dotplot(reorder(name, density) ~ 1000 * density, state.density,
          scales = list(x = list(log = 10)),
          xlab = "Density (per square mile)")
```

Taking logarithms alleviates the problem with Figure 10.15, and is generally preferable over artificial axis breaks. The inline call to `reorder()` creates a new factor with the same values and the same levels as `name`, but the levels are now ordered such that the first level is associated with the lowest value of `density`, the second with the next lowest, and so on. In this example there is exactly one value of `density` associated with each level of `name`, but this will not be true in general. In the following call, we reorder the levels of the `region` variable, a geographical classification of the states, again by `density`.

```
> state.density$region <-
      with(state.density, reorder(region, density, median))
```

Because there are multiple states for every region, we need to summarize the corresponding densities before using them to determine an order for the regions. The default summary is the mean, but here we use the median instead by specifying a third argument to `reorder()`.

Our eventual goal is to use `region` as a conditioning variable in a dot plot similar to the last one, but with states grouped by region. To do so, we first need to ensure that the levels of `name` for states within a region are contiguous, as otherwise they would not be contiguous in the dot plot.[6] This is achieved simply by reordering their levels by `region`. We would also like the states to be ordered by `density` within `region`, so we end up with another call to `reorder()` nested within the first one (this works because the original order is retained in case of ties).

```
> state.density$name <-
      with(state.density,
           reorder(reorder(name, density), as.numeric(region)))
```

We need to convert the `region` values to their numeric codes as the step of averaging would otherwise cause an error. Finally, we can use these reordered variables to produce the dot plot in Figure 10.20, with `relation = "free"` for the y-axis to allow independent scales for the different regions:

```
> dotplot(name ~ 1000 * density | region, state.density,
          strip = FALSE, strip.left = TRUE, layout = c(1, 4),
          scales = list(x = list(log = 10),
                        y = list(relation = "free")),
          xlab = "Density (per square mile)")
```

This still leaves room for improvement; the panels all have the same physical height, but different numbers of states, resulting in an odd looking plot. We could rectify this by changing the heights of the panels, as we did for widths in Figure 10.16. A convenience function that does this automatically, by making

---

[6] Factors are simply converted to the underlying numeric codes when they are plotted, and the codes are defined by the order of their levels.

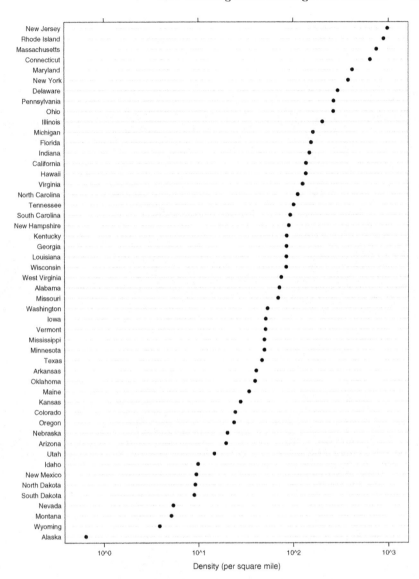

Figure 10.19. Population density in U.S. states, on a log scale. Comparing to Figure 10.15, we see that the states with the highest density no longer seem unusual compared to the rest. On the other hand, Alaska stands out as a state with unusually low density. Such judgments would have been harder if the states were not reordered.

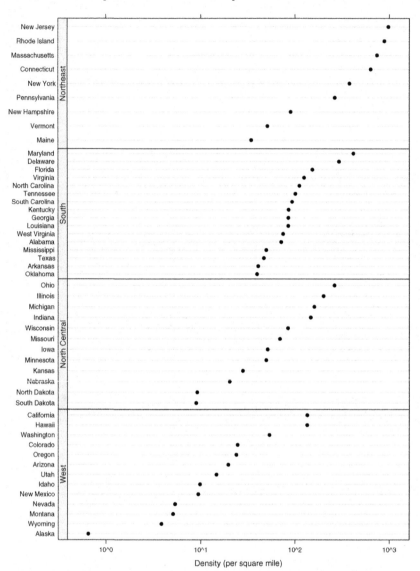

Figure 10.20. Population densities in U.S. states by region, with a different y-axis in each panel. Because the y variable is a factor, the scale calculations treat its values as the corresponding numeric codes; consequently, reordering of the levels is especially important. To appreciate this fact, the reader is strongly encouraged to try the call without reordering.

physical heights proportional to data ranges, is available in the latticeExtra package. Figure 10.21 is created by calling

```
> library("latticeExtra")
> resizePanels()
```

immediately after the previous dotplot() call. The inner workings of re-sizePanels() is explained in Chapter 12.

Another mode of automatic reordering is afforded by the index.cond argument, which we have used before without explanation to produce Figure 4.7. It implements a slightly different approach: it only reorders conditioning variables, and does so by trying to reorder packets based on their contents. To illustrate its use, consider the USCancerRates dataset in the latticeExtra package, which records average yearly deaths due to cancer in the United States between the years 1999 and 2003 at the county level. We plot the rates for men and women against each other conditioning by state, and order the panels by the median of the difference in the rates between men and women. Figure 10.22 is produced by

```
> data(USCancerRates, package = "latticeExtra")
> xyplot(rate.male ~ rate.female | state, USCancerRates,
         aspect = "iso", pch = ".", cex = 2,
         index.cond = function(x, y) { median(y - x, na.rm = TRUE) },
         scales = list(log = 2, at = c(75, 150, 300, 600)),
         panel = function(...) {
             panel.grid(h = -1, v = -1)
             panel.abline(0, 1)
             panel.xyplot(...)
         },
         xlab = "Deaths Due to Cancer Among Females (per 100,000)",
         ylab = "Deaths Due to Cancer Among Males (per 100,000)")
```

In general, index.cond can be a function, with a scalar numeric quantity as its return value, that is called with the same arguments as the panel function. When there is exactly one conditioning variable, its levels are reordered to put these return values in increasing order. For more than one conditioning variable, the order of each is determined by averaging over the rest.

10.7 Controlling the appearance of strips

Each panel in a multipanel lattice display represents a packet defined by a unique combination of levels of one or more conditioning variables. The purpose of the strips that typically appear above each panel is to indicate this combination. The contents of a strip can be customized using the strip argument. Strips can be placed to the left of each panel too, and these are controlled by the strip.left argument.

Both these arguments can be logical, with the corresponding strips suppressed if they are FALSE (strip.left is FALSE by default). The default

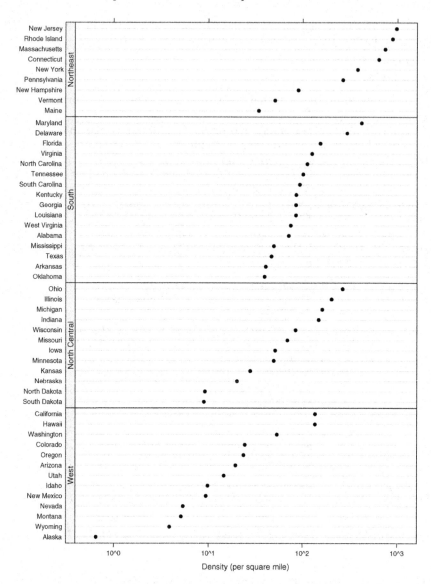

Figure 10.21. A variant of Figure 10.20, with panel heights varying by number of states.

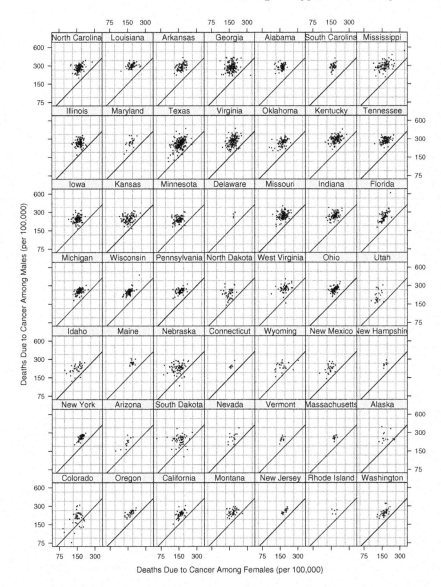

Figure 10.22. Annual death rates due to cancer (1999–2003) in U. S. counties by state for men and women, ordered by mean difference. A closer look reveals that the rate for women does not vary much across states, and the ordering is largely driven by the rate for men.

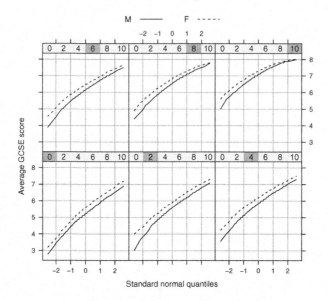

Figure 10.23. Normal Q–Q plots of average GCSE scores grouped by gender, conditioning on the A-level chemistry examination score, illustrating the use of a non-default strip annotation style. Another built-in style can be seen in Figure 11.5.

behavior can be changed by specifying the **strip** argument as a function that performs the rendering. One predefined function that can serve as a suitable strip function is **strip.default()**, which is used when **strip** is TRUE. **strip.default()** has several arguments that control its behavior, and these are often used to create variants. For example, one argument of **strip.default()** is **style**, which determines how the levels of a factor are displayed on the strip. We might define a new strip function as

```
> strip.style4 <- function(..., style) {
      strip.default(..., style = 4)
  }
```

When called, this function will call **strip.default()** with whatever arguments it received, with the exception of **style**, which will be changed to 4. This can be used to produce Figure 10.23 with

```
> data(Chem97, package = "mlmRev")
> qqmath(~gcsescore | factor(score), Chem97, groups = gender,
         type = c("l", "g"),  aspect = "xy",
         auto.key = list(points = FALSE, lines = TRUE, columns = 2),
         f.value = ppoints(100), strip = strip.style4,
         xlab = "Standard normal quantiles",
         ylab = "Average GCSE score")
```

Because it is common to create custom strip functions in this manner, a convenient generator function called strip.custom() is provided by lattice. strip.custom() is called with a subset of the arguments of strip.default() and produces another *function* that serves as a strip function by calling strip.default() after replacing the relevant arguments with their new values. Thus, an alternative call that produces Figure 10.23 is

```
> qqmath(~gcsescore | factor(score), Chem97, groups = gender,
          type = c("l", "g"), aspect = "xy",
          auto.key = list(points = FALSE, lines = TRUE, columns = 2),
          f.value = ppoints(100), strip = strip.custom(style = 4),
          xlab = "Standard normal quantiles",
          ylab = "Average GCSE score")
```

where the user-defined strip function is concisely specified inline. A full list of the arguments that can be manipulated in this manner (and their effect) is given in the help page for strip.default(). The above description applies to strip.left as well. Strips on the left can be useful when vertical space is at a premium, as in Figure 10.17. It is rarely useful to have both sets of strips (however, see Figure 11.6 and the accompanying discussion).

Another argument that indirectly controls the contents of the strip is par.strip.text, which is supplied directly to a high-level call. It is usually a list containing graphical parameters (such as col and cex) that are meant to control the appearance of the strip text. In practice, the list is passed on to the strip function, which may or may not honor it (the default strip function does). In addition to graphical parameters, par.strip.text can also contain a parameter called lines, which specifies the height of each strip in multiples of the default. The following example illustrates its use in conjunction with a custom strip function that does not depend on strip.default().

```
> strip.combined <-
      function(which.given, which.panel, factor.levels, ...) {
      if (which.given == 1) {
          panel.rect(0, 0, 1, 1, col = "grey90", border = 1)
          panel.text(x = 0, y = 0.5, pos = 4,
                     lab = factor.levels[which.panel[which.given]])
      }
      if (which.given == 2) {
          panel.text(x = 1, y = 0.5, pos = 2,
                     lab = factor.levels[which.panel[which.given]])
      }
  }
> qqmath(~ gcsescore | factor(score) + gender, Chem97,
          f.value = ppoints(100), type = c("l", "g"), aspect = "xy",
          strip = strip.combined, par.strip.text = list(lines = 0.5),
          xlab = "Standard normal quantiles",
          ylab = "Average GCSE score")
```

The lines component is used to halve the height of the strip, which would normally have occupied enough space for two strips. The actual strip function

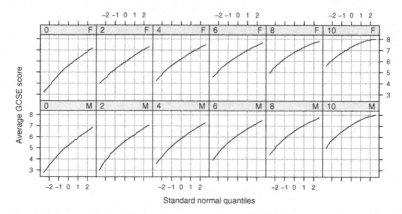

Figure 10.24. A variant of Figure 10.23, with gender now a conditioning variable as well, using a completely new strip function that incorporates both conditioning variables in a single strip.

is fairly transparent once we are told two facts: the strip functions for both conditioning variables use a common display area, and the native scale in the strip region is $[\,0, 1\,]$ in both axes. The resulting plot is shown in Figure 10.24.

10.8 An Example Revisited

Most of the examples we have seen in this book are designed to highlight some particular feature of lattice. Real life examples are typically more complex, requiring the use of many different features at once. Often, this simply results in a longer call. However, because of the way the various features of lattice interact, it is often possible to achieve fairly complex results with relatively innocuous looking code. To be able to write such code, one needs a familiarity with lattice that can only come from experience. We end this chapter with a study of one such example in some detail, with the hope that it will give the reader a sense of what can be achieved. The example is one we have encountered before; it was used to produce Figure 3.17:

```
> stripplot(sqrt(abs(residuals(lm(yield ~ variety+year+site)))) ~ site,
            data = barley, groups = year, jitter.data = TRUE,
            auto.key = list(points = TRUE, lines = TRUE, columns = 2),
            type = c("p", "a"), fun = median,
            ylab = expression(abs("Residual Barley Yield")^{1 / 2}))
```

The plot is based on the residuals from a linear model fit. Specifically, the square root of the absolute values of the residuals are plotted on the *y*-axis against one of the predictors (site) on the *x*-axis, with another predictor (year) used for grouping. The residuals are represented by points that are

jittered horizontally to alleviate overlap, and lines joining their medians are
added to summarize the overall trend. A legend describes the association
between the graphical parameters used to distinguish levels of **year** and the
actual levels.

To understand how we ended up with this call, let us consider how we
might approach the task of producing the plot given this verbal description.
The first step would be to fit an appropriate model, in this case

```
> fm <- lm(yield ~ variety + year + site, data = barley)
```

from which we could extract the residuals using

```
> residuals(fm)
```

Thus, the formula and data in the call might have been

```
> stripplot(sqrt(abs(residuals(fm))) ~ site, data = barley)
```

This is perfectly acceptable, but it runs the risk of a mismatch in the data
used in the model fit and the plot. We instead choose to incorporate the model
within the Trellis formula; the model is fit as part of the data evaluation step.

The next concern is the main display, which is controlled by the panel
function. In this case, the display should consist of the (jittered) points for
each group, along with a line joining the medians. Jittering is supported by the
default panel function **panel.stripplot()**, through the **jitter.data** argument. However, a look at the help page for **panel.stripplot()** indicates no
obvious way to add the lines, suggesting that we might need to write our own
panel function. The predefined panel function **panel.average()** is ideal for
our task; it even has an argument (**fun**) that can be used to specify the function that is used to compute the "average". Thus, our custom panel function
might look like

```
panel = function(x, y, jitter.data, ...) {
    panel.stripplot(x, y, jitter.data = TRUE, ...)
    panel.average(x, y, fun = median, ...)
}
```

Now, according to Table 5.1, **panel.average()** can be invoked through
panel.xyplot() by including "a" in the **type** argument. In addition, the
help page for **panel.stripplot()** notes that it passes all extra arguments to
panel.xyplot(). Add to this the fact that arguments unrecognized by **strip-
plot()** are passed along to the panel function, and we end up not requiring
an explicit panel function at all, as long as we add the suitable arguments
(**jitter.data**, **type**, and **fun**) to the high-level call. This also magically makes
the adjustments required to accommodate the **groups** argument. Of course,
such convenient panel functions are not always available, but they are often
enough to make this approach useful.

The final piece that completes the plot is the legend. We make use of the
auto.key argument, described in Chapter 9, which works by taking advantage
of the fact that the default plotting symbols and line types are obtained from
the global parameter settings. Since the output of **stripplot()** does not

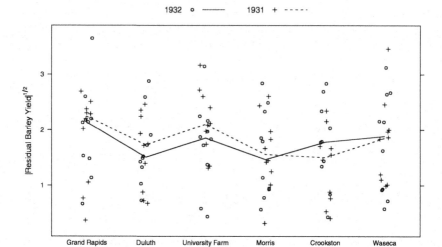

Figure 10.25. A spread–location plot of the barley data after "fixing" the Morris anomaly; compare with Figure 3.17.

normally contain lines, they are not included in the default legend, and we need to explicitly ask for them. Also notable is the use of `expression()` to specify the y-axis label. This is an example of the LATEX-like mathematical annotation (Murrell and Ihaka, 2000) that can be used in place of plain text in most situations.[7] In Figure 10.25, we redraw Figure 3.17, after switching the values of `year` for the observations from Morris using

```
> morris <- barley$site == "Morris"
> barley$year[morris] <-
      ifelse(barley$year[morris] == "1931", "1932", "1931")
```

The call to produce the plot remains otherwise unchanged.

---

[7] See `?plotmath` for a general description of these capabilities.

11

Manipulating the *"trellis"* Object

The Trellis paradigm is different from traditional R graphics in an important respect: high-level "plotting" functions in lattice produce objects rather than any actual graphics output. As with other objects in R, these objects can be assigned to variables, stored on disk in serialized form to be recovered in a later session, and otherwise manipulated in various ways. They can also be plotted, which is all we want to do in the vast majority of cases. Throughout this book, we have largely focused on this last task. In this chapter, we take a closer look at the implications of the object-based design and how one might take advantage of it.

11.1 Methods for *"trellis"* objects

The S language features its own version of object-oriented programming. To make things somewhat complicated, it has two versions of it: the *S3* or old-style version, and the newer, more formal *S4* version. The fundamental concepts are similar; objects have classes, and some functions are generic, with specific methods that determine the action of the generic when its arguments are objects of certain classes. However, the tools one can use to obtain information about a class or methods of a generic function are different. The lattice package is implemented using the *S3* system,[1] and the tools we describe in this section are specific to it.

The objects returned by high-level functions such as xyplot() have class *"trellis"*. We can obtain a list of methods that act specifically on *"trellis"* objects using

```
> methods(class = "trellis")

 [1] dimnames<-.trellis* dimnames.trellis*  dim.trellis*
 [4] plot.trellis*       print.trellis*     summary.trellis*
 [7] tmd.trellis*        [.trellis*         t.trellis*
```

---

[1] Although it is possible to extend it to *S4*, as we show in Chapter 14.

```
[10] update.trellis*
```

    Non-visible functions are asterisked

The output is useful primarily because it tells us where to look for documentation; for example, the documentation for the `dimnames()` method can be accessed by typing

```
> help("dimnames.trellis")
```

and that for the [method by typing

```
> help("[.trellis")
```

Note that this does not give a list of all generic functions that can act on *"trellis"* objects; for instance, `str()` is a generic function with no specific method for *"trellis"* objects, but a default method exists and that is used instead. These comments are not specific to the *"trellis"* class; for example, we could get a similar list of methods for *"shingle"* objects with

```
> methods(class = "shingle")
[1] as.data.frame.shingle* plot.shingle*
[3] print.shingle*          [.shingle*
[5] summary.shingle*
```

    Non-visible functions are asterisked

and a list for all methods for the generic function `barchart()` using

```
> methods(generic.function = "barchart")
[1] barchart.array*    barchart.default* barchart.formula*
[4] barchart.matrix*   barchart.numeric* barchart.table*
```

    Non-visible functions are asterisked

As hinted at by the output, most of these methods are not intended to be called by their full name. The correct usage is described in the respective help page, or sometimes in the help page for the generic. We now look at some of these methods in more detail.

11.2 The plot(), print(), and summary() methods

The most commonly used generic function in R is `print()`, as it is implicitly used to display the results of many top-level computations. For *"trellis"* objects, the `print()` method actually plots the object in a graphics device. It is sometimes necessary to use `print()`[2] explicitly, either because automatic printing would have been suppressed in some context, or to use one of the

---

[2] Or `plot()`, which is equivalent, except that it does not return a copy of the object being plotted.

optional arguments. The most useful arguments of the plot() and print() methods are described here briefly.

split, position

    These two arguments are used to specify the rectangular subregion within the whole plotting area that will be used to plot the *"trellis"* object. Normally the full region is used. The split argument, specified in the form c(col, row, ncol, nrow), divides up the region into ncol columns and nrow rows and places the plot in column col and row row (counting from the upper-left corner). The position argument can be of the form c(xmin, ymin, xmax, ymax), where c(xmin, ymin) gives the lower-left and c(xmax, ymax) the upper-right corner of the subregion, treating the full region as the $[0, 1] \times [0, 1]$ unit square.

more, newpage

    By default, a new "page" is started on the graphics device every time a *"trellis"* object is plotted. These two arguments suppress this behavior, allowing multiple plots to be placed together in a page. Specifying more = TRUE in a call causes the next *"trellis"* plot to be on the same page. Specifying newpage = FALSE causes the current plot to skip the move to a new page.[3]

panel.height, panel.width

    These two arguments allow control over the relative or absolute widths and heights of panels in terms of the very flexible unit system in grid. A full discussion of this system is beyond the scope of this book, but we show a simple example soon.

packet.panel

    This argument is a function that determines the association between packet order and panel order. The packet order arises from viewing a *"trellis"* object as an array with margins defined by the conditioning variables, with packets being the cells of the array. Just as regular arrays in R, this can be thought of as a vector with a dimension attribute, and the packet order is the linear order of packets in this vector. On the other hand, the panel order is the order of panels in the physical layout, obtained by varying the columns fastest, then the rows, and finally the pages. Specifying packet.panel allows us to change the default association rule, which is implemented by the packet.panel.default() function, whose help page gives further details and examples.

Other arguments of the plot() method are rarely needed and are not discussed here. Note that just as parameter settings normally specified using trellis.par.set() can be attached to individual *"trellis"* objects by adding a par.settings argument to high-level calls, arguments to the plot method can also be attached as a list specified as the plot.args argument.

---

[3] The latter is more general, as it allows lattice plots to be mixed with other grid graphics output. Specifically, newpage must be set to FALSE to draw a *"trellis"* plot in a previously defined viewport.

We have seen the use of the `plot()` method previously in Figures 1.4 (where the `split` and `newpage` arguments were used) and 10.11 (where `position` was used). In the next example, we illustrate the use of `more` to compare two common variants of the dot plot. The first step is to create variables representing suitable *"trellis"* objects.

```
> dp.uspe <-
      dotplot(t(USPersonalExpenditure),
              groups = FALSE,
              index.cond = function(x, y) median(x),
              layout = c(1, 5),
              type = c("p", "h"),
              xlab = "Expenditure (billion dollars)")
> dp.uspe.log <-
      dotplot(t(USPersonalExpenditure),
              groups = FALSE,
              index.cond = function(x, y) median(x),
              layout = c(1, 5),
              scales = list(x = list(log = 2)),
              xlab = "Expenditure (billion dollars)")
```

These are then plotted side by side in a 2 × 1 layout to produce Figure 11.1.

```
> plot(dp.uspe,     split = c(1, 1, 2, 1), more = TRUE)
> plot(dp.uspe.log, split = c(2, 1, 2, 1), more = FALSE)
```

Another useful method for *"trellis"* objects is `summary()`. For our next example, we create a dot plot similar to the one in Figure 10.21. The response this time is the `Frost` variable, which gives the mean number of days with minimum temperature below freezing between 1931 and 1960 in the capital or a large city in each U.S. state. We begin by defining a suitable data frame and then creating a *"trellis"* object

```
> state <- data.frame(state.x77, state.region, state.name)
> state$state.name <-
      with(state, reorder(reorder(state.name, Frost),
                          as.numeric(state.region)))
> dpfrost <-
      dotplot(state.name ~ Frost | reorder(state.region, Frost),
              data = state, layout = c(1, 4),
              scales = list(y = list(relation = "free")))
```

which we then summarize using the `summary()` method.

```
> summary(dpfrost)
Call:
dotplot(state.name ~ Frost | reorder(state.region, Frost), data = state,
    layout = c(1, 4), scales = list(y = list(relation = "free")))

Number of observations:
reorder(state.region, Frost)
```

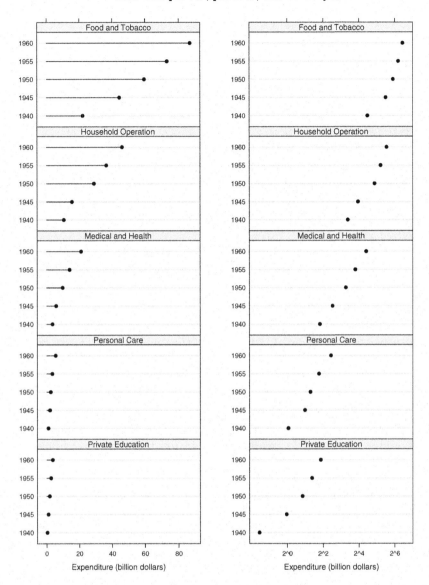

Figure 11.1. Two common variants of dot plots, showing trends in personal expenditure (on various categories) in the United States. Lines joining the points to a baseline, as in the plot on the left, are often helpful, but only if a meaningful baseline is available. In this case, patterns in the data are conveyed better by the plot on the right, with the data on a logarithmic scale.

```
      South         West     Northeast North Central
       16            13           9          12
```

The output gives us the call used to produce the object, but more important in this case, it gives us the number of observations (and hence, the approximate range of the y-axis) in each panel. We can use these frequencies to change the heights of the panels when plotting the object. Figure 11.2 is produced by[4]

```
> plot(dpfrost,
       panel.height = list(x = c(16, 13, 9, 12), unit = "null"))
```

This is not exactly what we want, as the actual range of the y-axis will be slightly different. However, the difference is practically negligible in this case. The `resizePanels()` function, used previously to produce Figure 10.21 and discussed further in the next chapter, does take the difference into account.

11.3 The `update()` method and `trellis.last.object()`

Perhaps the most useful method for *"trellis"* objects after `plot()` is `update()`, which can be used to incrementally change many (although not all) arguments defining a *"trellis"* object without actually recomputing the object. We have seen many uses of `update()` throughout this book and only give one more explicit example here.

`update()` is often useful in conjunction with the `trellis.last.object()` function. Every time a *"trellis"* object is plotted, whether explicitly or through implicit printing, a copy of the object is retained in an internal environment (unless this feature is explicitly disabled). The `trellis.last.object()` function can be used to retrieve the last object saved. Thus, the following command will produce Figure 11.3 when issued right after the previous example.

```
> update(trellis.last.object(), layout = c(1, 1))[2]
```

This example also illustrates the indexing of *"trellis"* objects as arrays. The above call recovers the last saved object using `trellis.last.object()` and updates it by changing the `layout` argument. Because the object had four packets, this would normally have resulted in a plot with four pages, but the indexing operator "[" is used to extract just the second packet.[5]

The indexing of *"trellis"* objects follows rules similar to those for regular arrays. In particular, indices can be repeated, causing packets to be repeated in the resulting plot. A useful demonstration of this feature is given in Figure 11.4, where a three-dimensional scatter plot with a single packet is displayed in multiple panels with gradually changing viewpoints. The figure is produced by

---

[4] The `"null"` unit is a special grid unit that asks the panels to be as tall as possible while retaining their relative heights. Other units such as `"inches"` or `"cm"` can be used to specify absolute heights.

[5] The `"["` method actually uses the `update()` method to change the `index.cond` argument, but is more intuitive and performs more error checks.

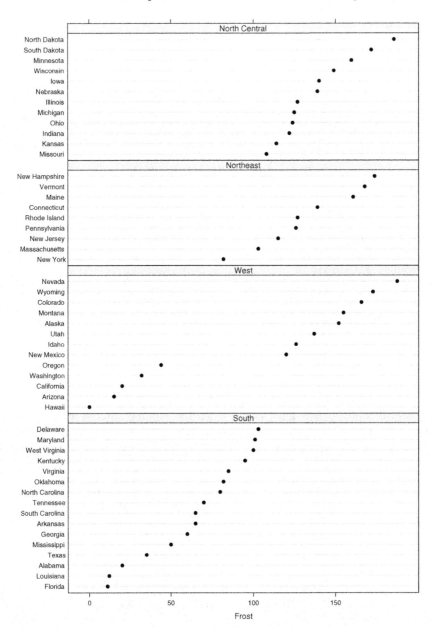

Figure 11.2. A dot plot similar to Figure 10.21, using the `Frost` column in the `state.x77` dataset. The heights of panels are controlled using a different method.

```
> npanel <- 12
> rot <- list(z = seq(0, 30, length = npanel),
              x = seq(0, -80, length = npanel))
> quakeLocs <-
      cloud(depth ~ long + lat, quakes, pch = ".", cex = 1.5,
            panel = function(..., screen) {
                pn <- panel.number()
                panel.cloud(..., screen = list(z = rot$z[pn],
                                               x = rot$x[pn]))
            },
            xlab = NULL, ylab = NULL, zlab = NULL,
            scales = list(draw = FALSE), zlim = c(690, 30),
            par.settings = list(axis.line = list(col="transparent")))
> quakeLocs[rep(1, npanel)]
```

The panel function makes use of the `panel.number()` function to detect which panel is currently being drawn. This and other useful accessor functions are described in Chapter 13.

11.4 Tukey mean–difference plot

The Tukey mean–difference plot applies to scatter plots and quantile plots. As the name suggests, the idea is to start with a set of (x, y) pairs, and plot the mean $(x + y)/2$ on the x-axis and the difference $x - y$ on the y-axis. In terms of plotting, this is equivalent to rotating the (x, y) data clockwise by 45°. The mean–difference plot is most useful when the original data lie approximately along the positive diagonal, as its purpose is to emphasize deviations from that line. M–A plots, popular in the analysis of microarray data, are essentially mean–difference plots.

It is fairly simple to create a mean–difference plot using `xyplot()` after manually transforming the data. As a convenience, the `tmd()` function performs this transformation automatically on *"trellis"* objects produced by `xyplot()`, `qqmath()`, and `qq()`. In the following example, we apply it to the two sample Q–Q plot seen in Figure 3.10. Figure 11.5, produced by

```
> data(Chem97, package = "mlmRev")
> ChemQQ <-
      qq(gender ~ gcsescore | factor(score), data = Chem97,
         f.value = ppoints(100), strip = strip.custom(style = 5))
> tmd(ChemQQ)
```

suggests that the distributions of `gcsescore` for girls and boys differ consistently in variance except for the lowest `score` group.

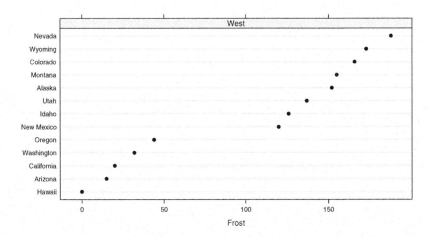

Figure 11.3. One panel from Figure 11.2, extracted from the underlying *"trellis"* object.

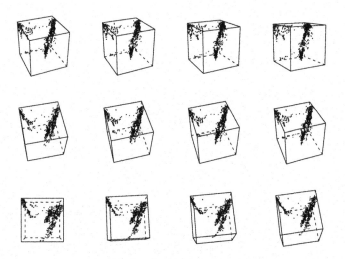

Figure 11.4. Varying camera position for a three-dimensional scatter plot of earthquake epicenter positions, from bottom left to top right.

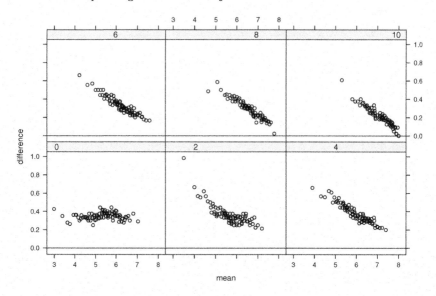

Figure 11.5. Tukey mean–difference plot, derived from the two-sample Q–Q plot in Figure 3.10.

11.5 Specialized manipulations

The use of strips on top of each panel to indicate levels of conditioning variables, introduced in the original Trellis implementation in S, was a remarkable innovation because it allowed multipanel displays with an arbitrary number of conditioning variables and a layout that is not necessarily tied to the dimensions of the conditioning variable.[6] This generality sometimes makes it difficult to implement designs that are perhaps more useful in special cases. For example, in a multipanel display with exactly two conditioning variables and the default layout (columns and rows representing levels of the first and second conditioning variables), one might want to indicate the levels only on the outer margins, once for each row and column, rather than in all panels. It is possible to realize such a design with lattice, but this requires far more detailed knowledge than warranted. Fortunately, the object model used in lattice makes it fairly simple to write functions that implement such specialized manipulations in a general way. In the next example, we make use of the use-OuterStrips() function in the latticeExtra package, which implements the design described above.

---

[6] In contrast, "conditioning plots" as previously implemented in the coplot() function indicated the association indirectly, and were limited to two conditioning variables.

Our example makes use of the `biocAccess` dataset, encountered previously in Figure 8.2. Here, we attempt to look at the pattern of access attempts over a day conditioned on month and day of the week.

```
> library("latticeExtra")
> data(biocAccess)
> baxy <- xyplot(log10(counts) ~ hour | month + weekday, biocAccess,
                 type = c("p", "a"), as.table = TRUE,
                 pch = ".", cex = 2, col.line = "black")
```

Just for fun, we note using the `dimnames()` method that the levels of the `month` variable are abbreviated month names, and change them to be the full names.

```
> dimnames(baxy)$month
[1] "Jan" "Feb" "Mar" "Apr" "May"
> dimnames(baxy)$month <- month.name[1:5]
> dimnames(baxy)

$month
[1] "January"  "February" "March"    "April"    "May"

$weekday
[1] "Monday"    "Tuesday"   "Wednesday" "Thursday"  "Friday"
[6] "Saturday"  "Sunday"
```

Of course, we could also have done this by writing a (fairly complicated) custom strip function, or more simply by modifying the levels of `month` beforehand. Finally, we call the `useOuterStrips()` function to produce a modified *"trellis"* object, which produces Figure 11.6.

```
> useOuterStrips(baxy)
```

Although not clear from this example, `useOuterStrips()` throws an error unless the requested manipulation is meaningful, and overrides any previously set layout.

11.6 Manipulating the display

Traditional R graphics encourages, and even depends on, an incremental approach to building graphs. For example, to create custom axis labels with traditional graphics, one would first create a plot omitting the axes altogether, and then use the `axis()` function, and perhaps the `box()` function, to annotate the axes manually. Trellis graphics, on the other hand, encourages the whole object paradigm, and the operation of updating serves as the conceptual analogue of incremental changes. The obvious advantage to this approach is that unlike traditional graphics, lattice displays can automatically allocate the space required for long axis labels, legends, and the like, because the labels or legends are known before plotting begins.

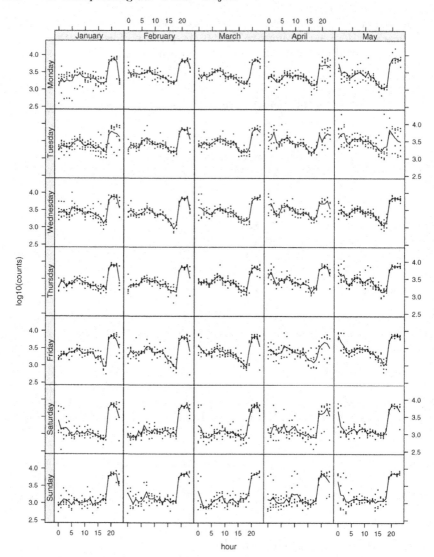

Figure 11.6. The number of hourly accesses to http://www.bioconductor.org, conditioning on month and day of the week. There is a difference in the patterns in weekdays and weekends, which can be seen more clearly in Figure 14.2. The strips have been manipulated so that they appear only on the top and the left of each column and row, rather than in each panel. This is a useful space-saving device when exactly two conditioning variables are used in the default layout. A color version of this plot is also available.

In some situations however, incremental additions are a necessary part of the workflow; for example, when one wants to identify and label certain "interesting" points in a scatter plot by clicking on them. This does not involve manipulation of the underlying object itself, but rather interaction with its visual rendering. The interface for such interaction is described in the next chapter.

12

Interacting with Trellis Displays

High-level functions in lattice produce *"trellis"* objects that can be thought of as abstract representations of visualizations. An actual rendering of a visualization is produced by plotting the corresponding object using the appropriate print() or plot() method. In this chapter, we discuss things the user can do after this plotting has been completed.

One possible approach is to treat the result as any other graphic created using the grid package, and make further enhancements to the display using the low-level tools available in grid. In particular, the display consists of a tree of viewports, and various grid graphical objects (grobs) drawn within them. The user can move down to any of these viewports and add further objects, or, less commonly, edit the properties of existing objects. The precise details of these operations are beyond the scope of this book, but are discussed by Murrell (2005). In this chapter, we focus entirely on a higher-level interface in the lattice package for similar tasks, which is less flexible,[1] but usually sufficient. The playwith package (Andrews, 2007) provides a user-friendly GUI wrapper for many of these facilities.

12.1 The traditional graphics model

In the traditional R graphics model, displays are often built incrementally. An initial plot is created using a high-level function (such as boxplot()), and further commands, such as lines() and axis(), add more elements to the existing display. This approach works because there is exactly one figure region, and there is no ambiguity regarding which coordinate system is to be used for additional elements. Things are not as simple in a multipanel Trellis display, as one needs the additional step of determining to which panel further increments should apply.

---

[1] In particular, it provides no facilities for editing existing graphical objects in the manner of grid.edit().

The recommended approach in the Trellis system is to encode the display using the panel function. This ties in neatly with the idea of separating control over different elements of a display; in this paradigm, the panel function represents the procedure that visually encodes the data. In some ways, this takes the incremental approach to the extreme; a panel starts with a blank canvas, with only the coordinate system set up, and the panel function is responsible for everything drawn on it. An apparent deficiency of this model is that the only "data" available to the panel function is the packet produced by the conditioning process. In practice, further data can be passed in through the ... argument, and panel-specific parts can be extracted if necessary using the subscripts mechanism and accessor functions such as packet.number() and which.packet() (see Chapter 13). A more real deficiency is that this paradigm does not include any reasonable model for interaction.

12.1.1 Interaction

Native R graphics has rather limited support for interaction, but what it does have is often useful. The primary user-level functions related to interaction in traditional R graphics are locator() and identify(). locator() is a low-level tool, returning locations of mouse clicks, and identify() is a slightly more specialized function that is used to add text labels to a plot interactively.

The grid analogue of locator() is grid.locator(), which returns the location of a single mouse click in relation to the currently active viewport. lattice uses grid.locator() to provide a largely API-compatible analogue of identify() called panel.identify(), along with a couple of other similar functions. However, before we can illustrate the use of these functions, we need some more background on the implementation of lattice displays.

12.2 Viewports, trellis.vpname(), and trellis.focus()

An elementary understanding of grid viewports is necessary to appreciate the API for interacting with lattice plots. Viewports are essentially arbitrary rectangular regions inside which plotting can take place. For our purposes, their most important feature is that they have an associated coordinate system.[2] The process of plotting a "trellis" object involves the creation of a number of viewports; for example, every panel, strip, and label has its own viewport. These viewports are retained after plotting is finished, and the associated viewport tree (showing the nesting of viewports within other viewports) can be displayed by calling

```
> library("grid")
> current.vpTree()
```

---

[2] Points in this coordinate system can be represented in a variety of units, see ?unit in the grid package for details.

To add to the display in a particular viewport (usually one corresponding to a panel), we need to first make it the active viewport.

Every viewport has a name that can be used to revisit it (using the grid functions `downViewport()` and `seekViewport()`). To make the viewport names predictable, lattice uses the function `trellis.vpname()` to create the relevant names. For example, the names of the x-label viewport and the strip viewport at column 2 and row 1 might be

```
> trellis.vpname("xlab", prefix = "plot1")
[1] "plot1.xlab.vp"
> trellis.vpname("strip", column = 2, row = 1, prefix = "plot2")
[1] "plot2.strip.2.1.vp"
```

where the `prefix` argument is a character string that potentially allows viewports for multiple *"trellis"* displays on a page to be distinguished from each other. However, the user does not typically need to know this level of detail and can instead use the functions `trellis.focus()` and `trellis.unfocus()` to navigate the viewport tree.

The viewport that is active after a *"trellis"* object has been plotted is the one in which the plotting started (this is usually the root viewport that covers the entire device). The `trellis.focus()` function is used to make a different viewport in the viewport tree active. For example, the panel viewport at column 2 and row 1 might be selected by calling

```
> trellis.focus("panel", column = 2, row = 1)
```

Most arguments of `trellis.vpname()` can be supplied to `trellis.focus()` directly. In addition, it checks for invalid `column` and `row` values and gives an informative error message if necessary. More important, it makes the most common uses slightly simpler. With a single panel display, simply calling

```
> trellis.focus()
```

with no arguments selects the panel. For a multipanel display (on an interactive screen device), the same call allows the user to choose a panel by clicking on it. The viewport chosen by `trellis.focus()` is highlighted by default, making it easy to identify for further interaction. Many of these details can be controlled by additional arguments to `trellis.focus()`. Finally, calling

```
> trellis.unfocus()
```

reverts to the original viewport after undoing any highlighting.

12.3 Interactive additions

Once the desired viewport is active, further additions can be made to the display by making suitable function calls. Such additions usually involve interaction. `grid.locator()` can be used to identify locations of individual mouse clicks, which then need to be handled appropriately. A typical use of

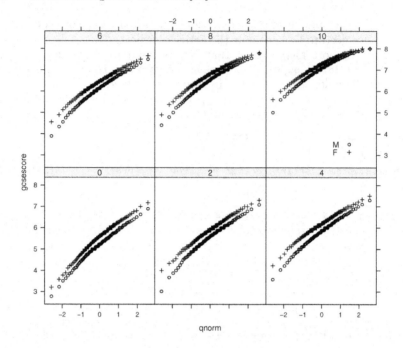

Figure 12.1. A normal quantile plot of `gcsescore` conditioning on `score` and grouping by `gender`. The legend describing the group symbols has been placed inside the plot interactively by clicking on the desired position.

this is to place a legend interactively on a plot. For example, the following code might produce Figure 12.1 after the user clicks on a suitable location.

```
> data(Chem97, package = "mlmRev")
> qqmath(~ gcsescore | factor(score), Chem97, groups = gender,
         f.value = function(n) ppoints(100),
         aspect = "xy",
         page = function(n) {
             cat("Click on plot to place legend", fill = TRUE)
             ll <- grid.locator(unit = "npc")
             if (!is.null(ll))
                 draw.key(simpleKey(levels(factor(Chem97$gender))),
                          vp = viewport(x = ll$x, y = ll$y),
                          draw = TRUE)
         })
```

In this example, the **page** argument has been used to encapsulate the process of asking for a user click and using the result to draw a suitable legend. The `draw.key()` function is normally used to construct a legend, as discussed in Chapter 9, but here it also draws the legend inside a newly created viewport.

The grid function `viewport()` is used to create the temporary viewport on the fly; the new viewport is centered on the location of the mouse click. We did not need to use `trellis.focus()` because we were not adding to any specific panel.

More complicated interaction modes can be built around `grid.locator()`. lattice provides three (at the time of writing) built-in functions that implement somewhat specialized forms of interaction. These are `panel.identify()`, `panel.identify.qqmath()`, and `panel.link.splom()`. We start with an illustration of `panel.identify()`, which is intended to be used with scatter plots as produced by `xyplot()` to add labels to points chosen interactively. Figure 12.2, showing a scatter plot of the murder rate against life expectancy in U.S. states with a few states labeled, might be the result of

```
> state <- data.frame(state.x77, state.region)
> xyplot(Murder ~ Life.Exp | state.region, data = state,
         layout = c(2, 2), type = c("p", "g"), subscripts = TRUE)
> while (!is.null(fp <- trellis.focus())) {
      if (fp$col > 0 & fp$row > 0)
          panel.identify(labels = rownames(state))
  }
```

There are several points that merit explanation in this sequence of calls. The first is the use of the `subscripts = TRUE` argument in `xyplot()` call. As noted in Section 5.2, panel functions can request an argument called `subscripts` that would contain the indices defining the rows of the data which form the packet in a given panel. Our intention is to label points using the corresponding state names, which are obtained from the row names of the `state` data frame. This represents names for all the data points, whereas we need names that correspond to the states used in individual panels. Obviously, `subscripts` gives us the right set of indices to extract the appropriate subset. Unfortunately, the subscripts are normally not retained if the panel function does not explicitly have an argument called `subscripts`. Specifying `subscripts = TRUE` in the high-level call forces retention of the subscripts, and is advisable for any call that is to be followed by interactive additions.

The next point of note is the use of `trellis.focus()` inside a `while()` loop. As mentioned earlier, calling `trellis.focus()` without arguments allows the user to select a panel interactively. Such a selection can be terminated by a right mouse button click (or by pressing the ESC key for the quartz device), in which case `trellis.focus()` returns NULL. We use this fact to repeatedly select panels until the user explicitly terminates the process in this manner. The user could also click outside the panel area, or on an empty panel; in this case, `trellis.focus()` returns a list with the row and col components set to 0 (for a normal selection, these would contain the location of the selection). We make sure that a valid selection has been made before we call `panel.identify()` to interactively label points inside the panel.

The final point is the use of `panel.identify()` every time a panel is successfully selected. When called, it allows the user to click on or near points in

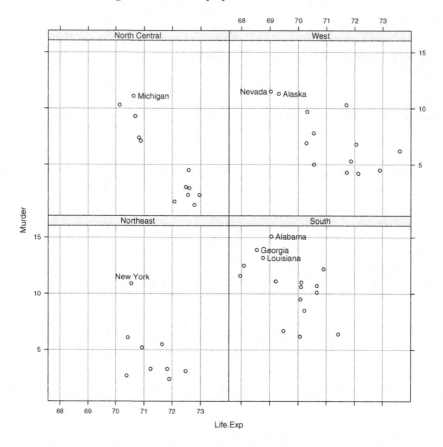

Figure 12.2. A scatter plot of murder rate versus life expectancy in U.S. states by region. In each panel, one or more states have been identified (labeled) by interactively selecting the corresponding points.

the selected panel to label them. This process continues until all points are labeled, or the process is explicitly terminated. Our call to `panel.identify()` specifies only one argument, `labels`, containing the labels associated with the full dataset. To make use of these labels, `panel.identify()` also needs to know the coordinates of the data points associated with these labels, and possibly the subscripts that need to be applied before the association is made. These arguments can be supplied to it as the `x`, `y`, and `subscripts` arguments. When `panel.identify()` is called after a call to `trellis.focus()` (or inside the panel function), these arguments may be omitted; they default to the corresponding arguments that would have been supplied to the panel function. Thus, the appropriate choice is made in every panel without explicit involvement of the user. This automatic selection is made using the `trellis.panelArgs()` function, which in turn uses `trellis.last.object()` to retrieve the last *"trellis"* object plotted. The correct set of arguments is determined using the accessor function `packet.number()`. This and other similar accessor functions are described more formally in Chapter 13.

Our next example illustrates the use of `panel.identify.qqmath()`, which is designed to add labels to a quantile plot produced by `qqmath()`. Figure 12.3 is produced by (after the appropriate pointing and clicking by the user)

```
> qqmath(~ (1000 * Population / Area), state,
        ylab = "Population Density (per square mile)",
        xlab = "Standard Normal Quantiles",
        scales = list(y = list(log = TRUE, at = 10^(0:3))))
> trellis.focus()
> do.call(panel.qqmathline, trellis.panelArgs())
> panel.identify.qqmath(labels = row.names(state))
> trellis.unfocus()
```

Most of the remarks concerning the previous example also apply here. Because the display has only one panel, calling `trellis.focus()` selects it automatically, and no interaction is required. An interesting addition is the call to `panel.qqmathline()`, through `do.call()`, which causes a reference line to be added as if `panel.qqmathline()` had been called as part of the panel function. This time, the correct panel arguments need to be retrieved explicitly using `trellis.panelArgs()`. This approach allows us to make incremental additions to individual panels of a lattice display, much as with the traditional graphics model. This facility is sometimes useful, although its regular use is not recommended as it detracts from the ideal of the *"trellis"* object as an abstraction of the entire graphic.

Our last example of interaction involves the `panel.link.splom()` function, which is designed to work with displays produced by `splom()`. When called, the user can click on a point in any of the subpanels to highlight the corresponding observation in all subpanels. Figure 12.4 is produced by

```
> env <- environmental
> env$ozone <- env$ozone^(1/3)
> splom(env, pscales = 0, col = "grey")
> trellis.focus("panel", 1, 1, highlight = FALSE)
> panel.link.splom(pch = 16, col = "black")
> trellis.unfocus()
```

The `trellis.focus()` call here explicitly chooses a panel, removing any possibility of interaction (although this is redundant in this case as there is only one panel). In addition, setting `highlight = FALSE` ensures that no decoration is added; without it, the display would have been redrawn when the call to `trellis.unfocus()` removed the decoration.

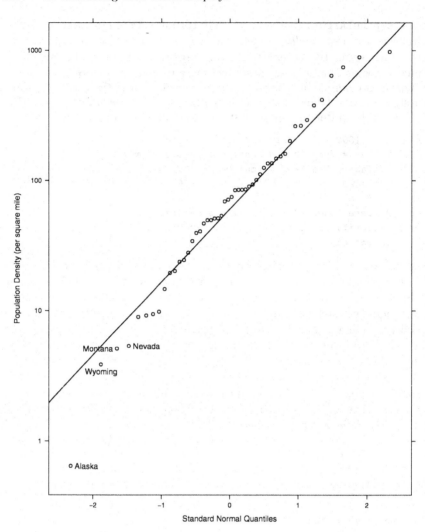

Figure 12.3. Normal quantile plot of population density in U.S. states. Some states have been labeled interactively after adding a reference line through the first and third quartile pairs.

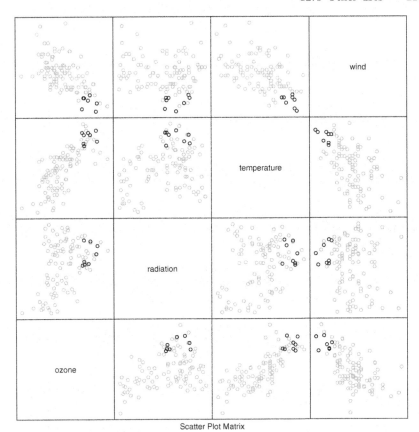

Scatter Plot Matrix

Figure 12.4. Interaction with a scatter-plot matrix. Clicking on a point highlights the corresponding observation in all subpanels.

12.4 Other uses

As we have already seen, it is possible to add pieces to a lattice display non-interactively after it has been drawn. Such use is often convenient, although the same effect can usually be achieved with a suitable panel function. Often, it is useful to simply interrogate a display to obtain information that is not easily available otherwise. For example, consider Figure 11.2, which is a dot plot of the mean number of days with minimum temperature below freezing in the capital or a large city in each U.S. state, conditioning on region. We reproduce the plot in Figure 12.5, but use the same height for every panel initially.

```
> state$name <- with(state,
                reorder(reorder(factor(rownames(state)), Frost),
                    as.numeric(state.region)))
```

```
> dotplot(name ~ Frost | reorder(state.region, Frost), data = state,
          layout = c(1, 4), scales = list(y = list(relation="free")))
```

Now that the graphic has been plotted, we can obtain the physical layout of panels in the display using the `trellis.currentLayout()` function (see Chapter 13)

```
> trellis.currentLayout()
     [,1]
[1,]    1
[2,]    2
[3,]    3
[4,]    4
```

and use it to compute the exact height of each panel in its native coordinate system:

```
> heights <-
      sapply(seq_len(nrow(trellis.currentLayout())),
             function(i) {
                 trellis.focus("panel", column = 1, row = i,
                               highlight = FALSE)
                 h <- diff(current.panel.limits()$ylim)
                 trellis.unfocus()
                 h
             })
> heights
[1] 16.2 13.2  9.2 12.2
```

It is now trivial to redraw the plot with the physical heights of the panels exactly proportional to their native heights, as was the intent of Figure 11.2. The following produces Figure 12.6.

```
> update(trellis.last.object(),
         par.settings = list(layout.heights = list(panel = heights)))
```

The `resizePanels()` function in the latticeExtra package, used for the same purpose to produce Figure 10.21, is simply this algorithm implemented with some sanity checks.

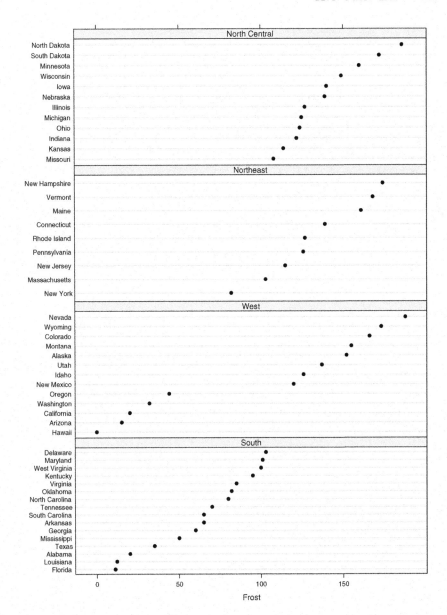

Figure 12.5. Redisplay of Figure 11.2, showing average number of days below freezing in U.S. states, conditioned on geographical region. Each panel has a different number of states, but the same physical height.

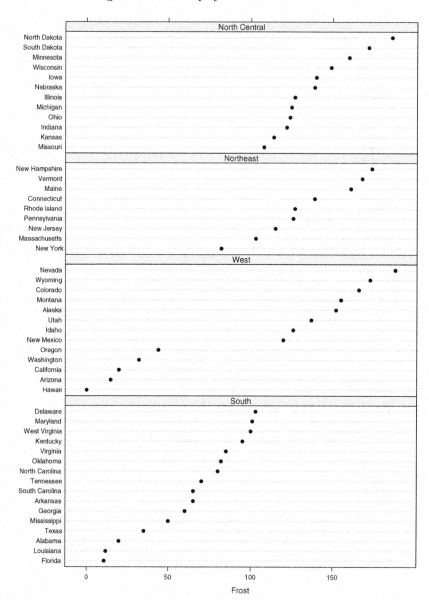

Figure 12.6. Updated form of Figure 12.6, with the physical heights of panels exactly proportional to native heights.

Part III

Extending Trellis Displays

13

Advanced Panel Functions

R is a complete programming language that allows, and indeed encourages, its users to go beyond the canned uses built into the system. The transition from user to programmer can be intimidating for the beginner to contemplate, but is almost inevitable after a point. In the context of lattice, this transition is most often necessitated by a desire to customize the display in small ways, perhaps just to add a common reference line to all panels. Such customizations are fairly basic in any serious use of lattice, and we have seen a number of examples throughout this book. In this chapter, which is meant for the more advanced user, we take a more formal look at panel functions, give pointers to the tools that might help in writing new ones, and finally discuss some nontrivial examples.

13.1 Preliminaries

Panel functions are like any other R function, except that they are expected to produce some graphical output when they are executed. They typically do so by calling a series of simpler panel functions, which might be viewed as building blocks of the complete display. Often, one also needs to manipulate the data available to the panel function before encoding them in the display. In this section, we describe the simple low-level panel functions available for use as building blocks, as well as some other related utilities. We demonstrate their use in creating data-driven displays in the subsequent sections.

13.1.1 Building blocks for panel functions

As we noted in Chapter 12, lattice is implemented using the low-level tools in the grid package. This has two important implications in the context of panel functions. First, lattice panel functions can make full use of grid primitives such as `grid.points()` and `grid.text()` and all their features. Second, lattice panel functions cannot make use of traditional graphics primitives[1] such as `points()` and `text()`. The plotting actions performed by a lattice panel function can consist entirely of grid function calls; in fact, grid primitives are more flexible than their traditional counterparts. A full discussion of grid is beyond the scope of this book, but a detailed exposition can be found in Murrell (2005) and the online documentation accompanying the grid package.

For those already familiar with traditional graphics, one practical drawback of grid is that it has an incompatible interface; that is, to reimplement a `text()` call in grid, one cannot simply change the name of the function to `grid.text()`; one also needs to modify the argument list. This can be a nuisance particularly when writing code that is intended for use both in R and S-PLUS (the latter does not have an implementation of grid). To make life easier, lattice provides analogues of several traditional graphics primitives; these are implemented using grid, but are intended to be drop-in replacements for the corresponding traditional graphics functions. For example,

`panel.points()` draws points (or lines, depending on the `type` argument) with an argument list similar to that of `points()`,

`panel.lines()` is analogous to `lines()`, and draws lines joining specified data points,

`panel.text()` is like `text()` and adds simple text or LaTeX-like expressions,

`panel.rect()` draws rectangles like `rect()`,

`panel.polygon()` draws polygons like `polygon()`,

`panel.segments()` draws line segments like `segments()`, and

`panel.arrows()` draws arrows like `arrows()`, with a slightly more general interface.

Needless to say, these functions are less flexible than the underlying grid functions, particularly in the choice of coordinate system. The lattice package also provides several "utility" panel functions that are not quite as generic, but are primarily intended for inclusion in other panel functions rather than for use by themselves. Among these are

`panel.fill()`, which fills the panel with a given color,

`panel.grid()`, which draws a reference grid,

`panel.abline()`, which draws reference lines of various kinds,

`panel.curve()`, which draws a curve defined by a mathematical expression, like `curve()`,

`panel.mathdensity()`, which draws a probability density function,

`panel.rug()`, which draws "rugs" like `rug()`,

`panel.loess()`, which adds a LOESS smooth of the supplied data,

`panel.lmline()`, which adds a regression line fit to the data,

`panel.qqmathline()`, which adds a line through two quantiles of the data and a theoretical distribution, and is primarily useful with `qqmath()`,

---

[1] This is not entirely true. See Figure 14.5.

panel.violin(), which draws violin plots, a useful alternative to box-and-whisker plots, and

panel.average(), which draws lines after aggregating and summarizing one variable by the unique levels of another.

Finally, each high-level function has its own panel function that can be reused in other contexts; these include panel.bwplot() and panel.xyplot(), among others. The panel.superpose() function is particularly useful for superposed displays. It conveniently handles separation of graphical parameters and allows another function to be specified as the panel.groups argument; this function is used as the panel function for each group and is supplied the appropriate graphical parameters.

We make no attempt to describe each of these functions in detail, as that would make this book longer than it already is. Instead, we refer the reader to their respective help pages.

13.1.2 Accessor functions

In principle, panel functions require no information beyond the data that are to be graphically encoded in that panel. In particular, it should not need to know where in the physical layout the current panel is located, nor should it worry about whether the current axis limits are appropriate for the data being encoded; it is expected that an appropriate data rectangle (viewport in grid jargon) with a suitable coordinate system has already been set up, and an appropriate clipping policy put in place, before the panel function is called. In practice, however, knowledge of these details can be important. Rather than supply such information through additional arguments, lattice provides a system of accessor functions that report the current state of the affairs when called from inside the panel function (or the strip or axis functions).

current.panel.limits() reports the limits of the current panel (viewport), typically in the native coordinate system, but possibly in any of the other systems supported by grid.

packet.number() returns an integer indicating which packet is being drawn. Packets are counted according to packet order, which is determined by varying the first conditioning variable the fastest, then the second, and so on.

panel.number() returns an integer counting which panel is being drawn, starting from one for the first panel. This is usually the same as the packet number, but not necessarily so.

trellis.currentLayout() returns a matrix with the same dimensions as the current layout of panels. The elements of the matrix indicate which packet (or panel) belongs in which position. For empty positions, the corresponding entry is 0.

current.row(), current.column() return the row or column position of the current panel in the layout.

which.packet() returns an integer vector as long as the number of conditioning variables, with each element an integer giving the current level of the corresponding variable.

These functions can be used while a *"trellis"* object is being plotted, as well as afterwards, while interacting with the display using the interface described in Chapter 12. As before, we refer the reader to the online documentation for more details.

13.1.3 Arguments

Panel functions are somewhat unusual in that they are rarely called by the user directly; they are instead called during the process of displaying a *"trellis"* object. This means, among other things, that the arguments available to a panel function are fully determined only in that context (recall that arguments supplied to a high-level function and not recognized by it are passed on to the panel function). To write a generally useful panel function, the author must take this fact into account. The arguments available will also depend on the relevant high-level function; for example, a panel function for xyplot() will expect arguments named x and y containing data, whereas one for densityplot() will only expect x. Usually, the most effective way to find out what arguments will be available (and how they should be interpreted) is to consult the help page of the default panel function, for example, panel.densityplot() for densityplot(). Of course, the most reliable way is to have the arguments listed explicitly; for example, using the panel function

```
> panel.showArgs <- function(...) str(list(...))
```

which is a function that simply writes out a compact summary of all its arguments. Not all potential arguments available to a panel function are necessarily supplied to it; only the ones that match the formal argument list of the panel function do, unless ... is one of the formal arguments. It is generally good practice to have a ... argument in panel functions and pass it on to further plotting functions, as this provides a simple mechanism to propagate graphical parameters.

One special argument in lattice panel functions is subscripts. If a panel function has a formal argument named subscripts, it will be called with subscripts containing the integer indices representing the rows in the original data (before any effect of subset) that define the packet used in that panel. Examples demonstrating the use of subscripts can be found in Section 5.2 and Chapter 12.

13.2 A toy example: Hypotrochoids and hypocycloids

Hypotrochoids are geometric curves traced out by a point within a circle that is rolling along "inside" another fixed circle. (Technically, the fixed circle can be smaller, in which case the moving circle is physically outside it.) They are examples of a more general class of curves called roulettes, which are generated by one object rolling along another. Hypotrochoids can be parameterized by the equations

$$x(t) = (R - r)\cos t + d\cos(R - r)\frac{t}{r}$$
$$y(t) = (R - r)\sin t - d\sin(R - r)\frac{t}{r}$$

where R is the radius of the fixed circle, r the radius of the moving circle, and d is the distance of the point being traced from the center of the latter. We can write a panel function that traces out this curve (with R fixed at 1) as follows.

```
> panel.hypotrochoid <- function(r, d, cycles = 10, density = 30)
  {
      if (missing(r)) r <- runif(1, 0.25, 0.75)
      if (missing(d)) d <- runif(1, 0.25 * r, r)
      t <- 2 * pi * seq(0, cycles, by = 1/density)
      x <- (1 - r) * cos(t) + d * cos((1 - r) * t / r)
      y <- (1 - r) * sin(t) - d * sin((1 - r) * t / r)
      panel.lines(x, y)
  }
```

This function has two interesting features; first, it does *not* have a ... argument, and second, none of the arguments is essential; even r and d are chosen randomly if they are missing. We show the implications of this in a moment.

First however, we consider hypocycloids, which are hypotrochoids with $d = r$, that is, the point being traced lying on the boundary of the moving circle. Hypocycloids are usually defined in terms of $k = 1/r$, and are closed curves when k is rational, with p "corners" if k is expressed as a ratio of two coprime integers p/q. We can write a simple wrapper function that draws hypocycloids as

```
> panel.hypocycloid <- function(x, y, cycles = x, density = 30) {
      panel.hypotrochoid(r = x / y, d = x / y,
                         cycles = cycles, density = density)
  }
```

where x and y represent q and p. We also need a prepanel function that defines the rectangle needed to fully contain a circle with unit radius centered at the origin:

```
> prepanel.hypocycloid <- function(x, y) {
      list(xlim = c(-1, 1), ylim = c(-1, 1))
  }
```

We can use the following code, producing Figure 13.1, to create a series of hypocycloids by varying the value of p while keeping q fixed.

Figure 13.1. Hypocycloids with parameter k varying from $11/10, 12/10, \ldots, 30/10$.

```
> grid <- data.frame(p = 11:30, q = 10)
> grid$k <- with(grid, factor(p / q))
> xyplot(p ~ q | k, grid, aspect = 1, scales = list(draw = FALSE),
        prepanel = prepanel.hypocycloid, panel = panel.hypocycloid)
```

This example is somewhat unusual in that the panel function is only provided two scalars at a time, which are used to compute and render a complex curve on the fly. Our next example is a whole lot more unusual. Figure 13.2 is produced by

```
> p <- xyplot(c(-1, 1) ~ c(-1, 1), aspect = 1, cycles = 15,
              scales = list(draw = FALSE), xlab = "", ylab = "",
              panel = panel.hypotrochoid)
> p[rep(1, 54)]
```

The panel function, `panel.hypotrochoid()`, does not accept arguments called x and y. Consequently, the x and y data specified in the formula do not get passed to the panel function at all; their sole purpose is to set up the data rectangle, avoiding the need for a prepanel function. In fact, the only argument explicitly passed on to the panel function is `cycles`, which determines the range of t that defines the curve. Thus, every time the panel function is called, a randomly chosen hypotrochoid is drawn. We draw several of them at once by repeating the first packet several times.

13.3 Some more examples

13.3.1 An alternative density estimate

As a more serious example, consider the problem of density estimation. The `densityplot()` function computes and displays density estimates given raw data, but it is restricted to the kernel density estimation methods implemented in the `density()` function. Suppose that we wish instead to use the log-spline density estimate (Stone et al., 1997) implemented in the logspline package (Kooperberg, 2007). Because the tools to compute the estimate are already available, writing a panel function to display it is fairly simple. To make sure our panels have the right height, we also have to write a suitable prepanel function.

```
> library("logspline")
> prepanel.ls <- function(x, n = 50, ...) {
      fit <- logspline(x)
      xx <- do.breaks(range(x), n)
      yy <- dlogspline(xx, fit)
      list(ylim = c(0, max(yy)))
  }
> panel.ls <- function(x, n = 50, ...) {
      fit <- logspline(x)
      xx <- do.breaks(range(x), n)
      yy <- dlogspline(xx, fit)
      panel.lines(xx, yy, ...)
  }
```

We can now use these to produce Figure 13.3 with

```
> faithful$Eruptions <- equal.count(faithful$eruptions, 4)
> densityplot(~ waiting | Eruptions, data = faithful,
              prepanel = prepanel.ls, panel = panel.ls)
```

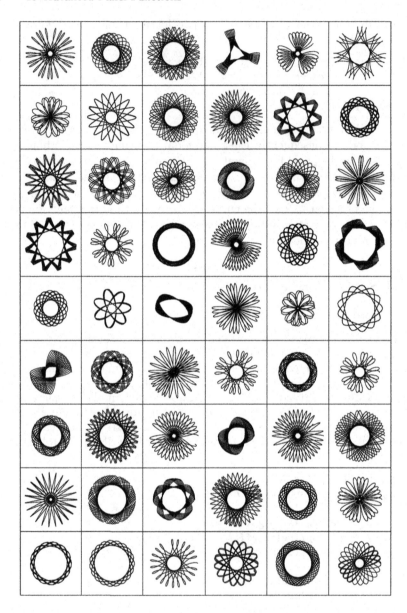

Figure 13.2. A series of hypotrochoids with randomly chosen parameters, reminiscent of the popular toy Spirograph®.

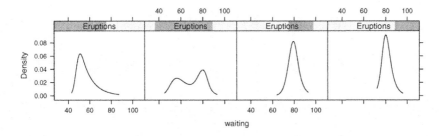

Figure 13.3. Conditional log-spline density estimates of waiting times before eruptions of the Old Faithful geyser, implemented with user-defined prepanel and panel functions.

13.3.2 A modified box-and-whisker plot

The next example is inspired by Tufte (2001), who describes a few variants of box-and-whisker plots that are motivated by the goal of reducing "non-data ink". In particular, the design we consider graphically summarizes the distribution of a continuous variable using a dot located at the median, and a couple of line segments extending from the first and third quartiles to the corresponding "extremes"; in other words, it is a box-and-whisker plot without the "box" (see `?boxplot.stats` for more concrete definitions). Our intention is not to comment on the merits of the design (especially because it is used here somewhat out of context), but simply to illustrate its implementation. A simple implementation is given by

```
> panel.bwtufte <- function(x, y, coef = 1.5, ...) {
      x <- as.numeric(x); y <- as.numeric(y)
      ux <- sort(unique(x))
      blist <- tapply(y, factor(x, levels = ux), boxplot.stats,
                      coef = coef, do.out = FALSE)
      blist.stats <- t(sapply(blist, "[[", "stats"))
      blist.out <- lapply(blist, "[[", "out")
      panel.points(y = blist.stats[, 3], x = ux, pch = 16, ...)
      panel.segments(x0 = rep(ux, 2),
                     y0 = c(blist.stats[, 1], blist.stats[, 5]),
                     x1 = rep(ux, 2),
                     y1 = c(blist.stats[, 2], blist.stats[, 4]),
                     ...)
  }
```

It is simple in the sense that it does not deal with "outliers" beyond the extremes and only produces vertical plots, but it is good enough to produce Figure 13.4 with

```
> data(Chem97, package = "mlmRev")
> bwplot(gcsescore^2.34 ~ gender | factor(score), Chem97,
```

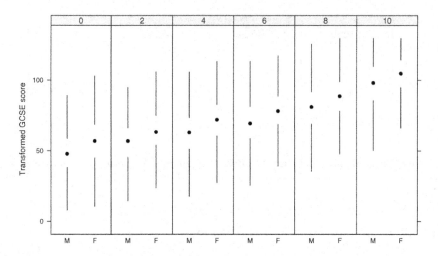

Figure 13.4. A variant of the standard box-and-whisker plot, showing the distribution of transformed GCSE scores by gender and the A-level chemistry examination score. The layout is the same as in Figure 3.12, but the encoding is inspired by an example from Tufte (2001).

```
panel = panel.bwtufte, layout = c(6, 1),
ylab = "Transformed GCSE score")
```

The result can be compared to Figure 3.12, which shows a regular box-and-whisker plot of the same data in the same layout.

13.3.3 Corrgrams as customized level plots

Corrgrams (Friendly, 2002) are visual representations of correlation matrices. They share the basic structure of a levelplot, but usually encode correlations by more than just color or grey level, and reorder the rows and columns by some measure of similarity. We continue with the example in Figure 6.13 to demonstrate a couple of variants.

```
> data(Cars93, package = "MASS")
> cor.Cars93 <-
      cor(Cars93[, !sapply(Cars93, is.factor)], use = "pair")
> ord <- order.dendrogram(as.dendrogram(hclust(dist(cor.Cars93))))
```

Our first panel function uses the ellipse package (Murdoch et al., 2007) to compute confidence ellipses representing correlation values, and additionally fills the ellipses with a color or grey level representing the correlation.

```
> panel.corrgram <-
      function(x, y, z, subscripts, at,
```

```
                    level = 0.9, label = FALSE, ...)
{
    require("ellipse", quietly = TRUE)
    x <- as.numeric(x)[subscripts]
    y <- as.numeric(y)[subscripts]
    z <- as.numeric(z)[subscripts]
    zcol <- level.colors(z, at = at, ...)
    for (i in seq(along = z)) {
        ell <- ellipse(z[i], level = level, npoints = 50,
                       scale = c(.2, .2), centre = c(x[i], y[i]))
        panel.polygon(ell, col = zcol[i], border = zcol[i], ...)
    }
    if (label)
        panel.text(x = x, y = y, lab = 100 * round(z, 2), cex = 0.8,
                   col = ifelse(z < 0, "white", "black"))
}
```

The panel function does not deal with colors explicitly, relegating that computation to the level.colors() function. Figure 13.5 is produced by

```
> levelplot(cor.Cars93[ord, ord], at = do.breaks(c(-1.01, 1.01), 20),
            xlab = NULL, ylab = NULL, colorkey = list(space = "top"),
            scales = list(x = list(rot = 90)),
            panel = panel.corrgram, label = TRUE)
```

Because there is no explicit color specification, the defaults provided by the theme active during plotting are used. Our second variant is similar, but this time uses partially filled circles to represent correlations. The circles are drawn using grid functions grid.polygon() and grid.circle() directly.

```
> panel.corrgram.2 <-
      function(x, y, z, subscripts, at = pretty(z), scale = 0.8, ...)
  {
      require("grid", quietly = TRUE)
      x <- as.numeric(x)[subscripts]
      y <- as.numeric(y)[subscripts]
      z <- as.numeric(z)[subscripts]
      zcol <- level.colors(z, at = at, ...)
      for (i in seq(along = z))
      {
          lims <- range(0, z[i])
          tval <- 2 * base::pi *
              seq(from = lims[1], to = lims[2], by = 0.01)
          grid.polygon(x = x[i] + .5 * scale * c(0, sin(tval)),
                       y = y[i] + .5 * scale * c(0, cos(tval)),
                       default.units = "native",
                       gp = gpar(fill = zcol[i]))
          grid.circle(x = x[i], y = y[i], r = .5 * scale,
                      default.units = "native")
      }
  }
```

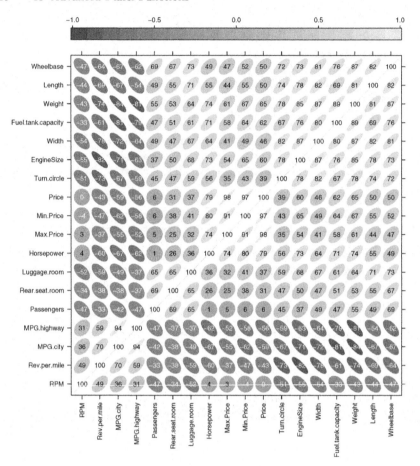

Figure 13.5. A corrgram, implemented as a `levelplot()` with a user-defined panel function, showing correlations using ellipses.

As before, color is not handled directly by the panel function. This time, however, we add a `col.regions` argument to the high-level call to `levelplot()`.

```
> levelplot(cor.Cars93[ord, ord], xlab = NULL, ylab = NULL,
           at = do.breaks(c(-1.01, 1.01), 101),
           panel = panel.corrgram.2,
           scales = list(x = list(rot = 90)),
           colorkey = list(space = "top"),
           col.regions = colorRampPalette(c("red", "white", "blue")))
```

col.regions is used by levelplot() to define the color key, but is also passed on to the panel function. From the perspective of the panel function, this is simply a part of the ... argument and is thus passed on unchanged to the level.colors() call, which does use it to compute suitable colors. The resulting display is shown with several other color plates in Figure 13.6.

13.4 Three-dimensional projections

Customizing the panel display in cloud() and wireframe(), the two high-level functions that make use of three-dimensional projection, is somewhat more involved. In addition to encoding the packet data, the panel function in this case is also responsible for drawing the bounding box and any axis annotation. One is usually interested only in changing the data-driven part of the display. This part can be controlled separately by specifying the panel.3d.cloud or panel.3d.wireframe arguments, which are technically arguments of the default panel function, and default to panel.3dscatter() and panel.3dwire(), respectively. More details can be found in the help page for these functions. Here, we give a simple example where a regular wireframe plot is supplemented by a contour plot "projected" onto the top surface of the bounding box. This involves computing the locations of the contour lines in the appropriate three-dimensional coordinate system, projecting it using ltransform3dto3d(), and drawing it. The following function executes these steps after calling panel.3dwire() to render the default wireframe display.

```
> panel.3d.contour <-
      function(x, y, z, rot.mat, distance,
              nlevels = 20, zlim.scaled, ...)
  {
      add.line <- trellis.par.get("add.line")
      panel.3dwire(x, y, z, rot.mat, distance,
                  zlim.scaled = zlim.scaled, ...)
      clines <-
          contourLines(x, y, matrix(z, nrow = length(x), byrow = TRUE),
                      nlevels = nlevels)
      for (ll in clines) {
          m <- ltransform3dto3d(rbind(ll$x, ll$y, zlim.scaled[2]),
                              rot.mat, distance)
          panel.lines(m[1,], m[2,], col = add.line$col,
                      lty = add.line$lty, lwd = add.line$lwd)
      }
  }
```

It can now be used in a call to wireframe() to produce Figure 13.7.

```
> wireframe(volcano, zlim = c(90, 250), nlevels = 10,
          aspect = c(61/87, .3), panel.aspect = 0.6,
          panel.3d.wireframe = "panel.3d.contour", shade = TRUE,
          screen = list(z = 20, x = -60))
```

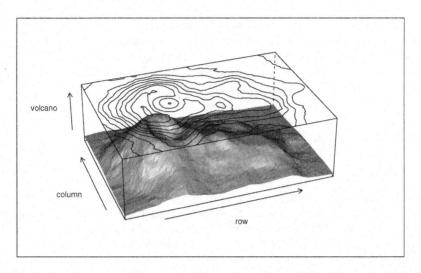

Figure 13.7. A three-dimensional view of the Maunga Whau volcano in Auckland created using `wireframe()`, with two-dimensional contours projected onto the top of the bounding box.

Figure 6.5 is another example of customized three-dimensional displays; the code to produce it is given later in this chapter.

It should be emphasized that the conventional R graphics model is far from optimal for three-dimensional displays; apart from the lack of dynamic manipulation, it has no high-level support for object occlusion, which makes it difficult to implement even the simplest of designs, such as a scatter plot combined with a fitted regression surface. Users who regularly work with three-dimensional displays should consider the **rggobi** and **rgl** packages, which provide interfaces to powerful alternative visualization systems.

13.5 Maps

Choropleth maps use color to encode a continuous or categorical variable on a map. Although somewhat specialized, choropleth maps are popular and effective in conveying spatial information. From an implementation perspective, there is nothing special about these plots; they are simply polygons with fill color derived from an external variable. The more important considerations are the practical ones of obtaining boundaries of the polygons defining the geographical units, and the associated data. In this section, we describe one approach that can be used to create choropleth maps using **lattice**, and point the reader to an alternative approach implemented in the **latticeExtra** package.

Tools to work with map data are available in the **maps** package (Becker et al., 2007) which contains, among other things, predefined boundary databases for several geographical units. In our examples, we use the "county" database, which contains information on counties in the United States. The map() function normally draws a map of a specified database, but can also be used to retrieve information about the polygons that define the map.

```
> library("maps")
> county.map <- map("county", plot = FALSE, fill = TRUE)
```

The fill argument causes the return value to be in a form that is suitable for use in polygon() (and hence panel.polygon()); it contains components x and y which are numeric vectors defining the boundaries, with NA values separating polygons. It also contains a vector of names for the polygons, which in this case represent U.S. counties.

```
> str(county.map)
List of 4
 $ x    : num [1:90997] -86.5 -86.5 -86.5 -86.6 ...
 $ y    : num [1:90997] 32.3 32.4 32.4 32.4 ...
 $ range: num [1:4] -124.7  -67.0    25.1    49.4
 $ names: chr [1:3082] "alabama,autauga" "alabama,baldwin" "alabama,..
 - attr(*, "class")= chr "map"
```

External data can be matched with polygons using these names. Getting the names into the same form may require some effort; we assume that this has already been done. Our first example uses the ancestry data in the latticeExtra package.

```
> data(ancestry, package = "latticeExtra")
> ancestry <- subset(ancestry, !duplicated(county))
> rownames(ancestry) <- ancestry$county
```

The data are derived from the U.S. 2000 census, and contain the most frequently reported ancestries in each county. As a first step, we pool the levels that appear infrequently.

```
> freq <- table(ancestry$top)
> keep <- names(freq)[freq > 10]
```

The row names of ancestry match the county names in county.map, and we use this fact to create a vector of ancestry values matching the map database.

```
> ancestry$mode <-
      with(ancestry,
           factor(ifelse(top %in% keep, top, "Other")))
> modal.ancestry <- ancestry[county.map$names, "mode"]
```

Finally, we use a color palette from the RColorBrewer package to produce Figure 13.8 (shown among the color plates). Thanks to the form of the value returned by map(), we can simply use panel.polygon() as our panel function, with a suitable vector of colors passed in as an argument to the high-level call.

```
> library("RColorBrewer")
> colors <- brewer.pal(n = nlevels(ancestry$mode), name = "Pastel1")
> xyplot(y ~ x, county.map, aspect = "iso",
          scales = list(draw = FALSE), xlab = "", ylab = "",
          par.settings = list(axis.line = list(col = "transparent")),
          col = colors[modal.ancestry], border = NA,
          panel = panel.polygon,
          key =
          list(text = list(levels(modal.ancestry), adj = 1),
               rectangles = list(col = colors),
               x = 1, y = 0, corner = c(1, 0)))
```

13.5.1 A simple projection scheme

Figure 13.8 plots county boundaries as if they lie on a plane, whereas they actually lie on a sphere. This is typically addressed by using one of several cartographic projection schemes, but another alternative is to convert the polygon boundaries into their three-dimensional representation, and use it in cloud(). This is demonstrated in the next example. First, we compute the coordinates of the respective polygons on the globe,

```
> rad <- function(x) { pi * x / 180 }
> county.map$xx <- with(county.map, cos(rad(x)) * cos(rad(y)))
> county.map$yy <- with(county.map, sin(rad(x)) * cos(rad(y)))
> county.map$zz <- with(county.map, sin(rad(y)))
```

and then define a panel function that draws polygons from three-dimensional data.

```
> panel.3dpoly <- function (x, y, z, rot.mat = diag(4), distance, ...)
  {
      m <- ltransform3dto3d(rbind(x, y, z), rot.mat, distance)
      panel.polygon(x = m[1, ], y = m[2, ], ...)
  }
```

Next, we use these to produce Figure 13.9 (see color plates).

```
> aspect <-
      with(county.map,
           c(diff(range(yy, na.rm = TRUE)),
             diff(range(zz, na.rm = TRUE))) /
           diff(range(xx, na.rm = TRUE)))
> cloud(zz ~ xx * yy, county.map, par.box = list(col = "grey"),
        aspect = aspect, panel.aspect = 0.6, lwd = 0.5,
        panel.3d.cloud = panel.3dpoly, col = colors[modal.ancestry],
        screen = list(z = 10, x = -30),
        key = list(text = list(levels(modal.ancestry), adj = 1),
                   rectangles = list(col = colors),
                   space = "top", columns = 4),
        scales = list(draw = FALSE), zoom = 1.1,
        xlab = "", ylab = "", zlab = "")
```

The `aspect` argument is required to ensure that the relative proportions of the x, y, and z scales are appropriate.

Another example of the use of maps in a three-dimensional display was given in Figure 6.5. We are finally in a position to understand the call that produced it. The critical step is to define a function that draws the state boundaries on the x–y plane.

```
> library("maps")
> state.map <- map("state", plot=FALSE, fill = FALSE)
> panel.3dmap <- function(..., rot.mat, distance, xlim, ylim, zlim,
                          xlim.scaled, ylim.scaled, zlim.scaled)
{
    scale.vals <- function(x, original, scaled) {
        scaled[1] + (x-original[1]) * diff(scaled) / diff(original)
    }
    scaled.map <- rbind(scale.vals(state.map$x, xlim, xlim.scaled),
                        scale.vals(state.map$y, ylim, ylim.scaled),
                        zlim.scaled[1])
    m <- ltransform3dto3d(scaled.map, rot.mat, distance)
    panel.lines(m[1,], m[2,], col = "grey76")
}
```

This is then used in combination with the default display to produce the desired effect.

```
> cloud(density ~ long + lat, state.info,
        subset = !(name %in% c("Alaska", "Hawaii")),
        panel.3d.cloud = function(...) {
            panel.3dmap(...)
            panel.3dscatter(...)
        },
        type = "h", scales = list(draw = FALSE), zoom = 1.1,
        xlim = state.map$range[1:2], ylim = state.map$range[3:4],
        xlab = NULL, ylab = NULL, zlab = NULL,
        aspect = c(diff(state.map$range[3:4]) /
                   diff(state.map$range[1:2]), 0.3),
        panel.aspect = 0.75, lwd = 2, screen = list(z = 30, x = -60),
        par.settings =
        list(axis.line = list(col = "transparent"),
             box.3d = list(col = "transparent")))
```

As in the previous example, much of the call is devoted to tweaking the aspect ratio and other such details.

13.5.2 Maps with conditioning

The use of `panel.polygon()` as the panel function does not work in multi-panel choropleth maps. The idea of using the map object as the data, with the actual variable of interest sneaked in as a color vector, is also somewhat artificial. A more natural approach is implemented by the `mapplot()` function in the latticeExtra package. We use it to obtain a multipanel choropleth map, this time visualizing a continuous response, the rate of death from cancer among males and females. The data are available in the USCancerRates dataset. The mapproj package (McIlroy et al., 2005) is used to apply a projection directly in the call to map(). Figure 13.10 (see color plates) is produced by

```
> library("latticeExtra")
> library("mapproj")
> data(USCancerRates)
> rng <- with(USCancerRates,
              range(rate.male, rate.female, finite = TRUE))
> nbreaks <- 50
> breaks <- exp(do.breaks(log(rng), nbreaks))
> mapplot(rownames(USCancerRates) ~ rate.male + rate.female,
          data = USCancerRates, breaks = breaks,
          map = map("county", plot = FALSE, fill = TRUE,
                    projection = "tetra"),
          scales = list(draw = FALSE), xlab = "",
          main = "Average yearly deaths due to cancer per 100000")
```

This example illustrates an important point, namely, that custom panel functions, although affording tremendous flexibility, are primarily useful in situations where the role of the variables involved fit into one of a few predefined models. In the next chapter, we discuss how to develop new high-level display functions, such as mapplot(), that let us bypass such constraints.

14

New Trellis Displays

Each high-level function in lattice is intended to create a certain type of statistical display by default. Many variations are already built into the default panel functions and can be activated with additional arguments in a high-level function call itself. More extensive modifications can be made by writing custom panel functions, as we have seen throughout this book and particularly in Chapter 13.

Although panel functions can be used to implement entirely novel visualizations, trying to shoehorn such a display into a function intended for another purpose is mostly useful as a one-off, quick-and-dirty solution. For a systematic implementation that could perhaps be used by others, it is often more sensible to create a new function whose name better reflects the nature of the visualization. On the other hand, existing function names are sometimes perfectly appropriate, and it is the data which are in a form that is not directly usable. A typical example of this is a univariate time series; there is really only one choice for the x and y variables in the xyplot() call that produced Figure 10.17, and the need for a new function to hide the use of a formula seems wasteful.

Rather than trying to anticipate all potential use cases, lattice provides the groundwork for further extensions by making use of the object-oriented features of R. Each high-level function in lattice is generic, with method dispatch possible on the first argument x and possibly (using the formal $S4$ system) the second argument data. New high-level display functions can be written either as new methods for existing generic functions, or, if it seems appropriate, as an entirely new function which should itself be generic to allow further specialized methods. In this chapter, we give examples of both new methods and new high-level functions implemented using the framework provided by lattice. These can, it is hoped, serve as models for further extensions.[1]

---

[1] Note that this is by no means the only way to extend lattice; the Hmisc and nlme packages are widely used examples that take different approaches.

14.1 *S3* methods

The high-level functions in lattice are generic functions, which means that new methods can be written to display objects based on their class. Such methods usually end up calling the corresponding *"formula"* method after some preliminary processing. They may have different defaults for some arguments, and even a few new ones. There are a few such methods built into lattice, such as histogram() and qqmath() methods for numeric vectors, levelplot() and wireframe() methods for matrices, and (somewhat nontrivial) barchart() and dotplot() methods for contingency tables as produced by table() or xtabs().

Here we give as examples two other methods, defined in the latticeExtra package, for the xyplot() generic. The first is for plotting time-series objects, and essentially performs the same task as the cutAndStack() function defined in Section 10.5.3. Figure 14.1 is produced by

```
> library("latticeExtra")
> xyplot(sunspot.year, aspect = "xy",
          strip = FALSE, strip.left = TRUE,
          cut = list(number = 4, overlap = 0.05))
```

This time, there is no need to write a wrapper function, and the cuts are specified using a new argument that is only meaningful for this method. Our second example, which is slightly more involved, is also related to time-series data. The stl() function decomposes a periodic time-series into seasonal, trend, and irregular components using LOESS (Cleveland et al., 1990). The result is an object of class *"stl"*; the xyplot() method for this class is used below to visualize the decomposition of the biocAccess data seen previously in Figure 8.2. The data are not in the form of a time-series, so we create one on the fly. To keep the plot from getting too compressed horizontally, data from only the first two months are used.

```
> data(biocAccess, package = "latticeExtra")
> ssd <- stl(ts(biocAccess$counts[1:(24 * 30 * 2)], frequency = 24),
             "periodic")
> xyplot(ssd, xlab = "Time (Days)")
```

The plot shows clear trends of decreased activity during weekends, as well as regular "seasonal" peaks of activity within each day (which happens to be caused by a poorly set up mirror).

Both these examples are primarily useful as demonstrations; *"stl"* objects have a plot() method that uses traditional graphics to produce an equivalent visualization, and the zoo package, which deals primarily with time-series data, has more general xyplot() methods for time-series objects. Other examples that can serve as prototypes are available in the coda package, and of course in the lattice package itself.

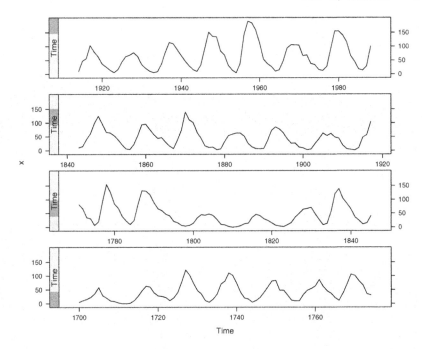

Figure 14.1. A cut-and-stack plot of the yearly number of sunspots between 1700 and 1988, created using an `xyplot()` method for time series data. The aspect ratio, chosen using the 45° banking rule, makes it easy to see that the ascent into peaks are usually steeper than the descents.

14.2 *S4* methods

Although the *S3* scheme works well for plotting highly structured objects, it is insufficient in situations where the flexibility of a formula interface is desirable, but with data objects that do not fit into the restrictive data frame paradigm.

This is important, for example, in the context of modern high-throughput bioinformatics data, where each "response" consists of thousands of measurements on the basic experimental unit, and covariate information on each experimental unit is stored as "phenotype data". The Bioconductor project (Gentleman et al., 2004) handles such data by defining new container classes. We can use such classes as alternative data sources in lattice methods using the multiple dispatch facilities in the *S4* system.[2] In Figure 14.3, we use the

---

[2] In the *S3* system, the specific method used when a generic function is called depends only on the class of one argument. *S4* generic functions, on the other hand, can select methods based on the classes of multiple arguments. This feature is known as multiple dispatch. The *S4* system has many other features not directly relevant for us; see Chambers (1998) for details.

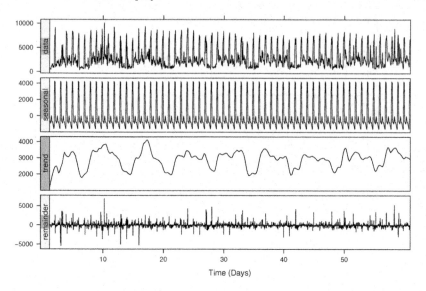

Figure 14.2. An STL decomposition of the hourly number of accesses to `http://www.bioconductor.org` over the period of two months, created using an `xyplot()` method for *"stl"* objects. The "trend" is periodic with a dip during weekends. The "seasonal" component shows the pattern of accesses over a day, with the spikes very likely due to automated activities such as mirroring.

`densityplot()` method from the Bioconductor package flowViz (Duong et al., 2007) that dispatches on a *"formula"* x and a *"flowSet"* data.

```
> library("flowViz")
> data(GvHD, package = "flowCore")
> densityplot(Visit ~ 'FSC-H' | Patient, data = GvHD)
```

The primary challenge in such examples is not multiple dispatch, but rather the handling of potentially large datasets. In this example, GvHD is a *"flowSet"* object containing data from 35 samples. Two of the variables in the formula (`Patient` ID and `Visit` number) represent phenotype data associated with the samples. Each sample produces a (on average) $15,000 \times 8$ data matrix; columns in these data matrices (e.g., FSC-H) are the variables we are interested in visualizing. The naïve approach would be to convert the full data into an expanded data frame (a "join" operation), but this would produce a data frame with roughly $15,000 \times 35$ rows! The solution used in the flowViz package is to use only the phenotype data to construct a lattice call; the actual data are stored in an environment (as part of the design of the *"flowSet"* class), and the panel and prepanel functions access only one sample at a time as necessary. The flowViz package contains several other examples of *S4* methods for high-level lattice functions.

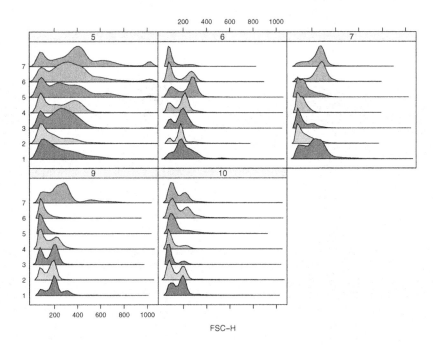

Figure 14.3. A visualization of the FSC-H channel in the GvHD data, created using a `densityplot()` method with signature (x = "formula", data="flowSet"). Each panel represents one patient, and the estimated densities of FSC-H for multiple visits are stacked on top of each other within each panel.

14.3 New functions

As we have already seen, existing generic function names may not be meaningful for new visualizations, and a completely new function name is often warranted. It is not necessary to define these functions as generic, but doing so has the benefit of encouraging future extensions. With a coordinated choice of argument names, it also allows multiple methods in multiple packages (perhaps written by different authors) to be used simultaneously without causing naming conflicts. We have already seen the `mapplot()` function in the lattice-Extra package used to produce Figure 13.10. Another prototypical example is the `hexbinplot()` function from the hexbin package, which is used as follows to produce Figure 14.4.

```
> library("hexbin")
> data(NHANES)
> hexbinplot(Hemoglobin ~ TIBC | Sex, data = NHANES, aspect = 0.8,
            trans = sqrt, inv = function(x) x^2)
```

The need to add an appropriate legend makes the implementation of `hexbinplot()` particularly instructive; the difficulty arises from the lack of a formal

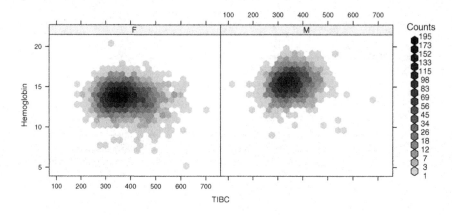

Figure 14.4. A conditional plot implementing the hexagonal binning algorithm of Carr et al. (1987), created using the **hexbinplot()** function. This example is somewhat challenging for the Trellis model, as it requires the panels to communicate information regarding bin counts to the legend.

mechanism to allow the panel function to communicate with the legend. The form of the legend itself poses another challenge, and requires nontrivial programming using grid. The interested reader is referred to the source code of the hexbin package for details.

14.3.1 A complete example: Multipanel pie charts

Care must be taken when writing new high-level functions to ensure that the expected nonstandard evaluation behavior is retained. Methods that call other high-level functions often need to delay the evaluation of certain arguments, and one way to do so is to make use of `match.call()` and `eval.parent()`. For our final example, we define a new high-level function that explicitly illustrates this approach.

The lattice package does not have a high-level function to draw pie charts because the information encoded by a pie chart can be conveyed more effectively by other graphs. They are a very familiar design nonetheless, and using the gridBase package (Murrell, 2005), which allows us to combine the normally incompatible traditional and grid graphics, we write a panel function that draws pie charts with minimal effort on our part:

```
> panel.piechart <-
      function(x, y, labels = as.character(y),
               edges = 200, radius = 0.8, clockwise = FALSE,
               init.angle = if(clockwise) 90 else 0,
               density = NULL, angle = 45,
```

```
                    col = superpose.polygon$col,
                    border = superpose.polygon$border,
                    lty = superpose.polygon$lty, ...)
{
    stopifnot(require("gridBase"))
    superpose.polygon <- trellis.par.get("superpose.polygon")
    opar <- par(no.readonly = TRUE)
    on.exit(par(opar))
    if (panel.number() > 1) par(new = TRUE)
    par(fig = gridFIG(), omi = c(0, 0, 0, 0), mai = c(0, 0, 0, 0))
    pie(as.numeric(x), labels = labels, edges = edges,
        radius = radius, clockwise = clockwise,
        init.angle = init.angle, angle = angle,
        density = density, col = col,
        border = border, lty = lty)
}
```

Because the form of data required by a pie chart is similar to that in a bar chart, we simply need to define a new function that calls barchart() with a new default panel function. Such a function is defined as

```
> piechart <- function(x, data = NULL, panel = "panel.piechart", ...)
{
    ocall <- sys.call(sys.parent())
    ocall[[1]] <- quote(piechart)
    ccall <- match.call()
    ccall$data <- data
    ccall$panel <- panel
    ccall$default.scales <- list(draw = FALSE)
    ccall[[1]] <- quote(lattice::barchart)
    ans <- eval.parent(ccall)
    ans$call <- ocall
    ans
}
```

Although this is not quite the standard way of writing functions in the S language, it ensures that arguments passed in as part of the ... argument of piechart() (which may include arguments such as groups and subset which follow special evaluation rules) are not evaluated prematurely. This function can now be used to produce Figure 14.5.

```
> par(new = TRUE)
> piechart(VADeaths, groups = FALSE, xlab = "")
```

We have ignored our own recommendation in not defining piechart() as a generic function, but this is easily fixed. Even as it stands, piechart() calls barchart() with minimal processing of its arguments, and consequently inherits the method dispatch behavior of barchart().

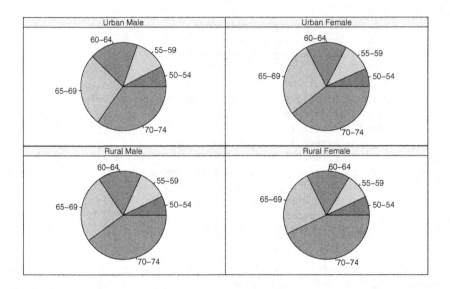

Figure 14.5. Conditional pie charts of the VADeaths data, created reusing the traditional graphics function pie() and the gridBase package. Compare with Figure 4.3, which presents the same data using a far more effective design.

References

F. Andrews. *playwith: A GUI for interactive plots using GTK+*, 2007. URL http://playwith.googlecode.com/. R package version 0.8.28.

A. Azzalini and A. W. Bowman. A look at some data on the Old Faithful Geyser. *Applied Statistics*, 39:357–365, 1990.

M. S. Bartlett. The square root transformation in analysis of variance. *Supplement to the Journal of the Royal Statistical Society*, 3(1):68–78, 1936.

R. A. Becker, A. R. Wilks, R. Brownrigg, and T. P. Minka. *maps: Draw Geographical Maps*, 2007. R package version 2.0-39.

G. E. P. Box and D. R. Cox. An analysis of transformations. *Journal of the Royal Statistical Society. Series B (Methodological)*, 26(2):211–252, 1964.

R. R. Brinkman, M. Gasparetto, S. J. J. Lee, A. J. Ribickas, J. Perkins, W. Janssen, R. Smiley, and C. Smith. High-content flow cytometry and temporal data analysis for defining a cellular signature of graft-versus-host disease. *BBMT*, 13(6):691–700, 2007.

S. M. Bruntz, W. S. Cleveland, B. Kleiner, and J. L. Warner. The dependence of ambient ozone on solar radiation, temperature, and mixing height. *Symposium on Atmospheric Diffusion and Air Pollution*, pages 125–128, 1974.

D. Carr, N. Lewin-Koh, and M. Maechler. *hexbin: Hexagonal Binning Routines*, 2006. R package version 1.12.0.

D. B. Carr, R. J. Littlefield, W. L. Nicholson, and J. S. Littlefield. Scatterplot matrix techniques for large *N*. *Journal of the American Statistical Association*, 82(398):424–436, 1987.

J. M. Chambers. *Programming with Data: A Guide to the S Language.* Springer, New York, 1998.

R. B. Cleveland, W. S. Cleveland, J. E. McRae, and I. Terpenning. STL: A seasonal-trend decomposition procedure based on loess. *Journal of Official Statistics*, 6:3–33, 1990.

W. S. Cleveland. *The Elements of Graphing Data.* Wadsworth, Monterey, California, 1985.

W. S. Cleveland, editor. *The Collected Works of John W. Tukey, Volume V: Graphics 1965–1985.* Wadsworth & Brooks/Cole Advanced Books & Software, Monterey, CA, 1988.

W. S. Cleveland. *Visualizing Data.* Hobart Press, Summit, New Jersey, 1993.

W. S. Cleveland and S. J. Devlin. Locally weighted regression: An approach to regression analysis by local fitting. *Journal of the American Statistical Association*, 83:596–610, 1988.

W. S. Cleveland and E. Grosse. Computational methods for local regression. *Statistics and Computing*, 1:47–62, 1991.

W. S. Cleveland, M. E. McGill, and R. McGill. The shape parameter of a two-variable graph. *Journal of the American Statistical Association*, 83: 289–300, 1988.

P. Dalgaard. *Introductory Statistics with R.* Springer, New York, 2002. URL http://www.biostat.ku.dk/~pd/ISwR.html. ISBN 0-387-95475-9.

T. Duong, B. Ellis, R. Gentleman, F. Hahne, N. Le Meur, and D. Sarkar. *flowViz: Visualization for flow cytometry*, 2007. R package version 1.3.0.

R. A. Fisher. *The Design of Experiments.* Hafner, New York, ninth edition, 1971.

M. Friendly. *Visualizing Categorical Data.* SAS Institute, Carey, NC, 2000. ISBN 1-58025-660-0.

M. Friendly. Corrgrams: Exploratory displays for correlation matrices. *The American Statistician*, 56(4):316–324, 2002.

R. C. Gentleman, V. J. Carey, D. M. Bates, B. Bolstad, M. Dettling, S. Dudoit, B. Ellis, L. Gautier, Y. Ge, J. Gentry, et al. Bioconductor: Open software development for computational biology and bioinformatics. *Genome Biology*, 5:R80, 2004. URL http://genomebiology.com/2004/5/10/R80.

W. Härdle. *Smoothing Techniques: With Implementation in S.* Springer, New York, 1990.

M. Harrower and C. A. Brewer. Colorbrewer.org: An online tool for selecting colour schemes for maps. *Cartographic Journal*, 40(1):27–37, 2003.

H. V. Henderson and P. F. Velleman. Building multiple regression models interactively. *Biometrics*, 37:391–411, 1981.

J. L. Hintze and R. D. Nelson. Violin plots: A box plot-density trace synergism. *The American Statistician*, 52:181–184, 1998.

R. Ihaka. Colour for presentation graphics. *Proceedings of DSC*, 2003. URL http://www.ci.tuwien.ac.at/Conferences/DSC-2003/Proceedings/Ihaka.pdf.

A. Inselberg. The plane with parallel coordinates. *The Visual Computer*, 1 (4):69–91, 1985.

W. B. Joyner and D. M. Boore. Peak horizontal acceleration and velocity from strong-motion records including records from the 1979 Imperial Valley, California, earthquake. *Bulletin of the Seismological Society of America*, 71 (6):2011–2038, 1981.

C. Kooperberg. *logspline: Logspline density estimation routines*, 2007. R package version 2.0.4.

C. Loader. *Local Regression and Likelihood.* Springer, New York, 1999.

R. Lock. 1993 New car data. *Journal of Statistics Education*, 1(1):7–7, 1993.

D. McIlroy, R. Brownrigg, and T. P. Minka. *mapproj: Map Projections*, 2005. R package version 1.1-7.1.

L. Molyneaux, S. K. Gilliam, and L. C. Florant. Differences in Virginia death rates by color, sex, age and rural or urban residence. *American Sociological Review*, 12(5):525–535, 1947.

D. Murdoch, E. D. Chow, and J. M. F. Celayeta. *ellipse: Functions for drawing ellipses and ellipse-like confidence regions*, 2007. R package version 0.3-5.

P. Murrell. *R Graphics*. Chapman & Hall/CRC, Boca Raton, FL, 2005. URL http://www.stat.auckland.ac.nz/~paul/RGraphics/rgraphics.html. ISBN 1-584-88486-X.

P. Murrell and R. Ihaka. An approach to providing mathematical annotation in plots. *Journal of Computational and Graphical Statistics*, 9(3):582–599, 2000.

R. B. Nelsen. *An Introduction to Copulas*. Springer, New York, 1999.

E. Neuwirth. *RColorBrewer: ColorBrewer palettes*, 2007. R package version 1.0-1.

R Development Core Team. *R: A Language and Environment for Statistical Computing*. R Foundation for Statistical Computing, Vienna, Austria, 2007. URL http://www.R-project.org. ISBN 3-900051-07-0.

J. Rasbash, F. Steele, W. Browne, and B. Prosser. *A User's Guide to MLwiN*. Institute of Education, University of London, 2000.

P. S. Reynolds. Time-series analyses of beaver body temperatures. In N. Lange, L. Ryan, L. Billard, D. Brillinger, L. Conquest, and J. Greenhouse, editors, *Case Studies in Biometry*, pages 211–228. Wiley-Interscience, 1994.

D. A. Rizzieri, L. P. Koh, G. D. Long, C. Gasparetto, K. M. Sullivan, M. Horwitz, J. Chute, C. Smith, J. Z. Gong, A. Lagoo, et al. Partially matched, nonmyeloablative allogeneic transplantation: Clinical outcomes and immune reconstitution. *Journal of Clinical Oncology*, 25(6), 2007.

D. W. Scott. Averaged shifted histograms: Effective nonparametric density estimators in several dimensions. *The Annals of Statistics*, 13:1024–1040, 1985.

C. J. Stone, M. H. Hansen, C. Kooperberg, and Y. K. Truong. Polynomial splines and their tensor products in extended linear modeling: 1994 Wald memorial lecture. *The Annals of Statistics*, 25(4):1371–1470, 1997.

D. F. Swayne, D. Temple Lang, A. Buja, and D. Cook. GGobi: Evolving from XGobi into an extensible framework for interactive data visualization. *Computational Statistics & Data Analysis*, 43(4):423–444, 2003.

E. R. Tufte. *The Visual Display of Quantitative Information*. Graphics Press, Cheshire, Connecticut, second edition, 2001.

J. W. Tukey. *Exploratory Data Analysis*. Addison-Wesley, Menlo Park, CA, 1977.

W. N. Venables and B. D. Ripley. *Modern Applied Statistics with S*. Springer, New York, fourth edition, 2002. URL http://www.stats.ox.ac.uk/pub/ MASS4. ISBN 0-387-95457-0.

E. J. Wegman. Hyperdimensional data analysis using parallel coordinates. *Journal of the American Statistical Association*, 85:664–675, 1990.

L. Wilkinson. *The Grammar of Graphics*. Springer, New York, 1999.

J. Yan and I. Kojadinovic. *copula: Multivariate Dependence with Copula*, 2007. R package version 0.5-8.

F. Yates. Complex experiments. *Journal of the Royal Statistical Society (Supplement)*, 2:181–247, 1935.

Index

Springer

the language of science

springer.com

Interactive and Dynamic Graphics For Data Analysis

Dianne Cook and Deborah F. Swayne

This richly illustrated book describes the use of interactive and dynamic graphics as part of multidimensional data analysis. Chapters include clustering, supervised classification, and working with missing values. A variety of plots and interaction methods are used in each analysis, often starting with brushing linked low-dimensional views and working up to manual manipulation of tours of several variables.

2007, Approx. 205 pp Softcover ISBN 978-0-387-71761-6

Graphics of Large Datasets Visualizing a Million

Antony Unwin, Martin Theus, and Heike Hoffman

This book shows how to look at ways of visualizing large datasets, whether large in numbers of cases, or large in numbers of variables, or large in both. All ideas are illustrated with displays from analyses of real datasets and the importance of interpreting displays effectively is emphasized. Graphics should be drawn to convey information and the book includes many insightful examples. The book is accessible to readers with some experience of drawing statistical graphics.

2006, XXII 271 pp. Hardcover ISBN 978-0-387-32906-2

Bayesian Computation with R

Antony Unwin, Martin Theus, and Heike Hoffman

This book introduces Bayesian modeling by the use of computation using the R language. Bayesian computational methods such as Laplace's method, rejection sampling, and the SIR algorithm are illustrated in the context of a random effects model. The construction and implementation of Markov Chain Monte Carlo (MCMC) methods is introduced. These simulation-based algorithms are implemented for a variety of Bayesian applications such as normal and binary response regression, hierarchical modeling, order-restricted inference, and robust modeling.

2007, X, 267 pp. Softcover ISBN 978-0—387-71384-7

Easy Ways to Order▶ Call: Toll-Free 1-800-SPRINGER • E-mail: orders-ny@springer.com • Write: Springer, Dept. S8113, PO Box 2485, Secaucus, NJ 07096-2485 • Visit: Your local scientific bookstore or urge your librarian to order.

CPSIA information can be obtained
at www.ICGtesting.com
Printed in the USA
LVOW05s1623230716

497508LV00001B/2/P